*Elements of Linear Algebra*

**A Series of Books in Mathematics**

*R. A. Rosenbaum and G. Philip Johnson*, EDITORS

# Elements of
# Linear Algebra

Daniel T. Finkbeiner II
*Kenyon College*

W. H. FREEMAN AND COMPANY
*San Francisco*

Printed in the United States of America

Library of Congress Catalog Card Number: 70–165416

International Standard Book Number: 0–7167–0445–5

1  2  3  4  5  6  7  8  9

# *Preface*

The study of linear algebra marks a significant stage in the mathematical development of a student—one that carries him beyond the familiar methods of algebra, geometry, and analysis to pave the way for the more sophisticated studies that characterize contemporary mathematics and many of its applications. Curricular trends during the past several decades have moved that stage earlier and earlier in the experience of the undergraduate. Thirty years ago a college course in linear algebra was rare indeed; ten years ago it was an upper level course for mathematics majors, rapidly gaining recognition as an essential part of the mathematical preparation of scientists, social scientists, and engineers. Today it is commonly offered as part of sophomore mathematics, following a year of calculus.

As the audience for linear algebra widens, pedagogical problems increase. To make substantial progress in the subject during a brief span of one quarter or one semester poses a genuine challenge to both the instructor and the student. Deliberate attention must be given to the selection of topics, the pace at which these topics unfold, and the depth to which they are pursued in the time available.

In a first course in linear algebra a student should learn some major facts about vector spaces, matrix algebra, systems of linear equations,

linear transformations, and quadratic forms. He should also learn some
basic techniques of computation. And he should develop the ability to
use language and notation precisely in reading and writing careful
mathematical proofs. To accomplish these aims, patience and self-
restraint are required of the instructor, and persistence is required of the
student.

To give the student an early feeling for the subject in the reasonably
familiar setting of three-dimensional Euclidean space, this book presents
most of the basic ideas of linear algebra informally in Chapter 1. The
same topics are studied again throughout the remaining chapters in a more
general setting, with more formality, and in greater detail. Although this
spiral approach might appear to the instructor to be an extravagant use
of time at the outset, experience in the classroom indicates that repetition
of ideas is important to the majority of students, who acquire mathematical maturity gradually. Familiarity with specific examples in three-
space generates valuable insight and confidence about analogous problems
in $n$-space.

The long introductory chapter contains a general survey of linear
algebra; the other four chapters provide substantial introductions to the
major topics of the subject: linear spaces (mostly real and finite-dimensional) in Chapter 2; linear mappings and matrix algebra in Chapter 3;
systems of linear equations and determinants in Chapter 4; characteristic
values and diagonalization problems in Chapter 5. In all, the book contains thirty-seven sections, some of which require more than one class
hour. Depending upon the length of the term and the ability of the class,
the instructor may find it advisable to omit or abbreviate Sections 1.13,
4.8, and all of Chapter 5. The first four chapters by themselves constitute
a respectable first course, and even Chapter 1 alone can be used as a
short survey course.

Students are strongly advised to make effective use of the exercises,
which include illustrative examples, problems for practice in computation, routine proofs, and theoretical results that extend the treatment of
the text and prepare for the development of subsequent theory. *All
exercises should be studied as part of the text*, whether or not they are
assigned for solution. Answers, hints, or detailed suggestions for almost
all exercises are provided at the end of the book to guide and assist the
student.

This book had its origin as a set of lecture notes written for the 1967
summer institute for high school teachers at Bowdoin College. The

generous response of those teacher-students encouraged me to extend those notes to a semester-length treatment that was used at Kenyon College for three years prior to publication. To all who contributed to the improvement of my account I owe unspoken gratitude. Particular acknowledgement is due to my daughter Susan, who typed the manuscript and provided critical insights as an undergraduate student of mathematics. The errors and flaws that remain are here in spite of these efforts. To Mary, my wife, I express a very special appreciation; understanding and forbearance do not produce a book, but they surely do help.

*April, 1971*                                              *Daniel T. Finkbeiner II*

# Contents

## 1
## *An Overview of Linear Algebra*

# 2
# Vector Spaces

# 3
# Linear Mappings and Matrices

# 4
# Systems of Linear Equations

# 5
# Diagonalization (Optional)

# *1*

# *An Overview of Linear Algebra*

Because this book is addressed principally to students whose mathematical preparation includes no more than a first course in calculus, this first chapter is intended to provide an informal survey of some of the major concepts and techniques of linear algebra in the concrete setting of three-dimensional Euclidean space and to develop a store of specific examples that will illustrate the more general theory developed in later chapters. Students who are already familiar with Euclidean three-space and who have already acquired reasonable facility with mathematical abstraction might be able to proceed quickly to Chapter 2. In general, however, the importance of acquiring a thorough mastery of the material in Chapter 1 should not be underestimated.

## 1.1  Introduction

The term *linear algebra* describes a general body of mathematics, closely related to geometry and analysis as well as to algebra itself, which has diverse applications in many subjects other than mathematics. The study centers on *linear* phenomena, and the principal methods of investigation are *algebraic*. A more precise understanding of these terms will be developed as our study progresses; for the moment it will suffice to recall that any line $L$ in the Euclidean plane can be expressed algebraically by a linear equation,

$$L: \quad ax + by = c,$$

where $a$, $b$, and $c$ are real numbers, and where $x$ and $y$ are formal symbols, sometimes called variables. If $x$ and $y$ are assigned numerical values, say $x = x_0$ and $y = y_0$, then point $P(x_0, y_0)$ lies on the line $L$ if and only if the value of the numerical expression

$$ax_0 + by_0$$

is the number $c$. It is reasonable to refer to the formal expression

$$ax + by$$

as a *linear combination* of $x$ and $y$, even when the symbols $x$ and $y$ might represent something other than numbers. For example, if $f$ and $g$ are real functions and if $a$ and $b$ are real numbers, the expression

$$af + bg$$

describes a real function $F$, defined by

$$F(x) = (af + bg)(x) = af(x) + bg(x)$$

for each real $x$ that is common to the domains of $f$ and $g$. Then $F$ is called a linear combination of $f$ and $g$ even though the graph of none of the functions $f$, $g$, or $F$ is a line.

   Another example of linearity, familiar from calculus, is the differentiation formula

$$D(af + bg) = aDf + bDg,$$

which states that the derivative of a given linear combination of functions is the same linear combination of the derivatives of those functions. In this case $D$ is said to be a *linear operator*.

   These examples illustrate briefly the concept of linearity; they also serve notice that linear algebra is concerned with various types of mathematical "objects" — numbers, points, functions, and operators. It is crucial that each type of object have the following closure property:

For any objects $\alpha$ and $\beta$ and any real numbers $a$ and $b$ the linear combination

$$a\alpha + b\beta$$

is an object of the same type.*

Moreover, we shall be interested in any linear function defined on such objects — that is, any function $G$ having the property that

$$G(a\alpha + b\beta) = aG(\alpha) + bG(\beta)$$

for all numbers $a$ and $b$ and all objects $\alpha$ and $\beta$. The precise nature of these objects is often a matter of indifference in the study of linear algebra; what does interest us is their behavior. In order to have a more frequently-used name for them than "objects," we shall call them *vectors*.

In this initial chapter we shall examine some of the important concepts of linear algebra within the special setting of three-dimensional Euclidean space to gain familiarity with basic ideas and methods that are applicable in a much more general context. Euclidean three-space provides an appropriate introduction to the subject because the geometry of this setting is familiar and general enough that an understanding of the concepts of linear algebra in three-space greatly assists in extending them to Euclidean $n$-space and to other vector spaces. Furthermore, three-space is a natural extension of the real line and the real plane, which we now review very briefly.

The system of real numbers will be denoted by $R$. If $x$ and $y$ are real numbers, then for any real numbers $a$ and $b$, $ax + by$ is also a real number. The arithmetic of real numbers supplies the properties that we shall later require for the type of algebraic system that is called a *vector space*. Hence $R$ is an example of a vector space: but as it is a bit too special to reveal the general nature of vector spaces we pass immediately to another example.

The set of all ordered *pairs* of real numbers will be denoted by $R \times R$. If $\alpha = (x, y)$ and $\beta = (u, v)$ are ordered pairs of real numbers, the sum $\alpha \oplus \beta$ is defined by

$$(x, y) \oplus (u, v) = (x + u, y + v),$$

*A list of the names and symbols of Greek letters appears on page 228.

**4**

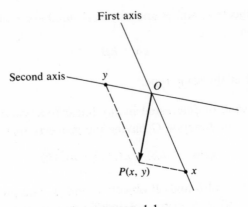

FIGURE 1.1

and the product of a real number $k$ and an ordered pair $\alpha = (x, y)$ is defined by

$$k \odot (x, y) = (kx, ky).$$

Special symbols $\oplus$ and $\odot$ are used here to emphasize that vector addition is not the same as addition of numbers, and that multiplication of a number and a vector is not the same as multiplication of two numbers. Vector algebra is concerned with two types of objects: an ordered pair $(x, y)$ of real numbers called a *vector*, and a single real number $k$ called a *scalar*. Although the usual arithmetic operations for real numbers apply to scalars, two special operations apply to vectors—vector addition and multiplication of a vector by a scalar coefficient.

When the set $R \times R$ is equipped with these operations, the resulting algebraic system is a vector space, which we denote by $R_2$.

As an example of a vector space, $R_2$ is more illustrative than $R$ because it emphasizes the distinction between vectors and scalars. In $R$ the real numbers serve both as vectors and scalars. From the definitions it is clear that for any $a$ and $b$ in $R$, the linear combination

$$(a \odot \alpha) \oplus (b \odot \beta)$$

is the ordered pair of real numbers, $(ax + bu, ay + bv)$—a member of $R_2$.

In order to give a geometric interpretation to the ordered pair $(x, y)$, we must first choose a coordinate system for the plane by selecting two intersecting coordinate axes, designating one as the first axis and the

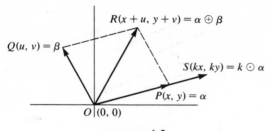

FIGURE 1.2

other as the second. Each ordered pair $(x, y)$ then determines that point $P$ of the coordinate plane having $x$ as its first coordinate and $y$ as its second coordinate. $P$ is then the terminal point of a directed line segment $OP$ whose initial point is the origin of that coordinate system, as indicated in Figure 1.1.

Often we shall need to use coordinate systems that are not rectangular, but because of its familiarity the rectangular case is convenient for simple illustrations. Unless the context indicates that another coordinate system is intended, a standard rectangular system may be used. Figure 1.2, for example, gives a geometric interpretation of the sum of two vectors and a scalar multiple of a vector relative to rectangular coordinates.

Thus we see that the word "vector" can have numerous meanings. In $R_2$ a vector $\alpha$ is simply an ordered pair $(x, y)$ of real numbers. Relative to a given coordinate system for the plane, $\alpha$ can be regarded geometrically as a point $P$ in the plane or as a directed line segment $OP$ from the origin to $P$. Moreover, there is the physical description of a vector as a quantity that has magnitude and direction; this describes a directed line segment not necessarily emanating from the origin. The geometric and physical concepts of a vector can be related by using the concept of congruent figures, familiar from geometry. Considering all directed line segments of the plane, we define two directed segments as congruent if and only if they have the same length and the same direction. Thus a directed segment $OP$ from the origin to $P$ is congruent to any directed segment (such as $QR$ in Figure 1.2) obtained by translating $OP$ parallel to itself in the plane.

Given any directed line segment (such as $ST$ in Figure 1.3) the class of *all* of its parallel translations is called a congruence class. Any two directed segments are congruent if and only if they are in the same con-

**6**

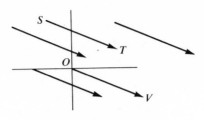

FIGURE - 1.3

gruence class, and each congruence class has one and only one of its members starting at the origin (*OV* in Figure 1.3). Viewed in this way, the set of all vectors of $R_2$ is seen to be in one-to-one correspondence with the set of all congruence classes of directed line segments of the coordinate plane.

A vector in $R_2$ can also be regarded as a translation of the points of the plane, but we choose not to pursue the idea except to emphasize that the term "vector" can be interpreted in various ways:

> *numerically*, as an ordered set of numbers,
> *physically*, as an arrow,
> *geometrically*, as a point *P* or directed segment *OP*.

It is especially important to keep in mind that the relation between numerical vectors and geometric vectors depends upon the coordinate system that is chosen for the geometric space. Fortunately, much of the study of linear algebra can be accomplished without the use of specified coordinate systems. We can often consider "vectors" as elements of an algebraic system whose rules of operations are known and are independent of any particular interpretation of the system. A vector regarded in this more abstract manner is called an *algebraic vector*.

## EXERCISES 1.1

1. In each case, write the vector (the directed line segment starting at (0, 0)) that is congruent to the directed line segment from *A* to *B*. State which of the directed segments are in the same congruence class.

> (i)  $A(0, 1)$,    $B(1, 3)$;
> (ii)  $A(-2, 1)$,    $B(-3, -1)$;
> (iii)  $A(3, -2)$,    $B(4, 0)$;

(iv)  $A(-3, 1)$,      $B(-1, 2)$;
(v)  $A(5, 4)$,       $B(3, 5)$;
(vi)  $A(-3, 1)$,      $B(-2, 3)$;
(vii)  $A(-5, 2)$,      $B(-3, 6)$.

2. Given $\alpha = (2, 3)$ and $\beta = (-1, 1)$, describe geometrically the locus of points corresponding to

(i)   the vectors $t \odot \alpha$, as $t$ varies over $R$;

(ii)  the vectors $\beta \oplus (t \odot \alpha)$, as $t$ varies over $R$;

(iii) the vectors $\beta \oplus (t \odot \alpha)$, as $t$ varies over the interval $-\infty < t \leq 1$;

(iv)  the vectors $(t \odot \alpha) \oplus (1 - t) \odot \beta$ as $t$ varies over the interval $0 \leq t \leq 1$.

3. If $B$ is the point corresponding to the vector $\beta = (b_1, b_2)$, if $\alpha = (a_1, a_2)$ is any vector other than $(0, 0)$, and if $\xi = (x, y)$ is an arbitrary vector, show that the vector equation

$$\xi = \beta \oplus (t \odot \alpha) \qquad \text{for all } t \text{ in } R,$$

gives a parametric representation for the line $L$ through $B$ and parallel to the vector $\alpha$. Write in parametric form two Cartesian equations that are equivalent to the single vector equation, and then eliminate the parameter $t$ to obtain a single Cartesian equation for the line $L$.

4. As in Exercise 3, describe the graph of the vector equation

$$\xi = (t \odot \alpha) \oplus ((1 - t) \odot \beta) \qquad \text{for all } t \text{ in } R.$$

Derive the corresponding parametric equations in Cartesian form and eliminate the parameter to obtain a familiar Cartesian equation for that locus.

5. If $\alpha$ and $\beta$ are vectors from the origin to points $A$ and $B$, indicate on a sketch the points that correspond to the vector $(t \odot \alpha) \oplus ((1 - t) \odot \beta)$ for the following values of $t$:

(i)   $t = \frac{1}{2}$;

(ii)  $t = 2$;

(iii) $t < 0$.

6. In $R_2$ the *scalar* or *dot product* of two vectors, $\alpha = (a_1, a_2)$ and $\beta = (b_1, b_2)$, is defined by $\alpha \cdot \beta = a_1 b_1 + a_2 b_2$. Use your knowledge of plane Euclidean geometry to derive the following relationships of the scalar product to the metric concepts of length and angle.

(i)   The length $\|\alpha\|$ of the vector $\alpha$ is $(\alpha \cdot \alpha)^{1/2}$.

(ii)  Nonzero vectors $\alpha$ and $\beta$ are perpendicular if and only if $\alpha \cdot \beta = 0$.

(iii) If $\Psi(\alpha, \beta)$ is the angle between nonzero vectors $\alpha$, $\beta$, then

$$\cos \Psi(\alpha, \beta) = \frac{\alpha \cdot \beta}{\|\alpha\| \, \|\beta\|}.$$

(iv) If $\alpha \neq (0, 0)$, the perpendicular projection of $\beta$ upon $\alpha$ is the vector

$$\frac{\alpha \cdot \beta}{\alpha \cdot \alpha} \odot \alpha.$$

(v) The area of the parallelogram having $\alpha$ and $\beta$ as adjacent sides is $[(\alpha \cdot \alpha)(\beta \cdot \beta) - (\alpha \cdot \beta)^2]^{1/2}$.

(vi) The result of (v) can also be expressed as $|a_1 b_2 - a_2 b_1|$. If you have studied determinants you will recognize this expression as the absolute value of a two-by-two determinant whose entries are the components of $\alpha$ and $\beta$.

7. Apply the information and methods described in Exercise 6 to

(i) show that the vectors described in Exercises 1(v) and (vi) are perpendicular;

(ii) determine the length of the vectors described in Exercises 1(iv) and (vi);

(iii) compute the area of the parallelogram having as adjacent sides the vectors described in Exercises 1(iv) and (vi).

8. Given the vectors $\alpha = (2, 1)$ and $\beta = (-3, 2)$. Use the information and methods described in Exercise 6 to compute

(i) $\alpha \cdot \beta$, $\|\alpha\|$, $\|\beta\|$,

(ii) the cosine of the angle between $\alpha$ and $\beta$,

(iii) the perpendicular projection of $\alpha$ upon $\beta$,

(iv) the area of the parallelogram having $\alpha$ and $\beta$ as adjacent sides.

## 1.2 The Vector Space $R_3$

Our first two examples of vector spaces were constructed from $R$ and $R \times R$, so it is not surprising that our next sample is constructed from $R \times R \times R$, the set of all ordered *triples* of real numbers. In order to make this set into a vector space we must define for $R_3$ two algebraic operations, vector sum and scalar multiple, to satisfy certain axioms yet to be specified. If $\alpha = (a_1, a_2, a_3)$ and $\beta = (b_1, b_2, b_3)$ are ordered triples of real numbers, and if $k$ is a real number, then *vector sum* and *scalar multiple* are defined by

$$\alpha \oplus \beta = (a_1, a_2, a_3) \oplus (b_1, b_2, b_3) = (a_1 + b_1, a_2 + b_2, a_3 + b_3),$$

$$k \odot \alpha = k \odot (a_1, a_2, a_3) = (ka_1, ka_2, ka_3).$$

In this example we note again that two different sets of objects are used in a vector space: the set of all ordered triples of real numbers, which are called *vectors*, and the set of all real numbers, which are called *scalars*. The two algebraic operations for $R_3$ imitate the corresponding operations for $R_2$: the sum of two ordered triples is defined component-by-component to yield another ordered triple; the scalar multiple of an ordered triple is also defined component-by-component to yield another ordered triple. Again, distinctive symbols denote these algebraic operations on triples, emphasizing that they are different from the corresponding operations on real numbers. Later, when there is no possibility of confusing the two pairs of operations, we shall dispense with the special symbols.

To interpret these operations geometrically we regard each ordered triple as a geometric vector by selecting a coordinate system for three-dimensional space. It is not necessary for the coordinate axes to be mutually perpendicular, but usually we choose this familiar coordinate system for ease of illustration. The numerical vectors $\alpha$ and $\beta$ correspond uniquely to points $A$ and $B$ that have the specified coordinates and thus correspond to geometric vectors $OA$ and $OB$ (see Figure 1.4). The vector $\alpha \oplus \beta$ corresponds to point $C$ that has coordinates $(a_1 + b_1, a_2 + b_2, a_3 + b_3)$ and thus to the geometric vector $OC$. It is not difficult to verify that if $O, A$, and $B$ are not collinear, then the four points $OACB$ are successive vertices of a parallelogram in the plane determined by the three points $O, A$, and $B$. If $O, A$, and $B$ are collinear, a similar interpretation can be made for the collapsed parallelogram.

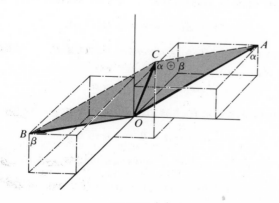

FIGURE 1.4

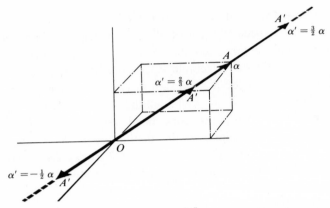

FIGURE 1.5

This interpretation describes the *parallelogram principle* of vector addition, which may already be familiar to you in terms of the resultant of two forces, interpreted as physical vectors.

The scalar multiple $k \odot \alpha$ of a vector corresponds to a point $A'$ collinear with $O$ and $A$. The Euclidean length of the line segment $OA'$ is $|k|$ times the length of $OA$. If $k > 1$, $A$ lies between $O$ and $A'$; if $0 < k < 1$, $A'$ lies between $O$ and $A$; if $k < 0$, $O$ lies between $A$ and $A'$. (See Figure 1.5.)

Direct computation with ordered triples will verify that the algebraic operations we have just defined satisfy the following properties.

(1)  Vector sum is *commutative:*    for all vectors $\alpha$, $\beta$ in $R_3$

$$\alpha \oplus \beta = \beta \oplus \alpha.$$

(2)  Vector sum is *associative:*    for all vectors $\alpha$, $\beta$, $\gamma$ in $R_3$

$$(\alpha \oplus \beta) \oplus \gamma = \alpha \oplus (\beta \oplus \gamma).$$

(3)  There exists in $R_3$ a *zero vector*, denoted $\theta$:    for each vector $\alpha$ in $R_3$

$$\alpha \oplus \theta = \alpha = \theta \oplus \alpha.$$

(4)  Each vector $\alpha$ in $R_3$ has an *additive inverse*, denoted $-\alpha$:

$$\alpha \oplus -\alpha = \theta = -\alpha \oplus \alpha.$$

(5)  For all scalars $a$, $b$ in $R$ and each vector $\alpha$ in $R_3$

$$(a + b) \odot \alpha = (a \odot \alpha) \oplus (b \odot \alpha).$$

(6)   For each scalar $a$ in $R$ and all vectors $\alpha$, $\beta$ in $R_3$

$$a \odot (\alpha \oplus \beta) = (a \odot \alpha) \oplus (a \odot \beta).$$

(7)   For all scalars $a$, $b$ in $R$ and each vector $\alpha$ in $R_3$

$$(ab) \odot \alpha = a \odot (b \odot \alpha).$$

(8)   For each vector $\alpha$ in $R_3$

$$1 \odot \alpha = \alpha.$$

In the next chapter we shall formally define a *vector space* as any algebraic system composed of vectors and scalars in which

(a)   the sum of any two vectors is a vector,

(b)   any scalar multiple of any vector is a vector,

(c)   the operations of vector sum and scalar multiple satisfy Properties (1–8).

Properties (2), (3), and (4) describe a general algebraic system called a *group*. Thus (1–4) can be expressed concisely by saying that the vectors of a vector space form a *commutative group* relative to the operation of vector sum. Properties (5–8) relate the operation of scalar multiple to the operation of vector sum. Properties (1–8) can be used easily to prove the following three additional properties for any vector space.

(9)   For each vector $\alpha$, $0 \odot \alpha = \theta$.

(10)   For each vector $\alpha$, $(-1) \odot \alpha = -\alpha$.

(11)   For each scalar $a$, $a \odot \theta = \theta$.

From this brief introduction it should be clear that linear algebra is concerned with various kinds of mathematical objects and different sets of objects of the same kind. It will greatly facilitate our work to adopt concise notation that will distinguish effectively between one kind of object and another. Therefore, we will denote *scalars* (real numbers) by lower case Latin letters and *vectors* by lower case Greek letters. (A Greek alphabet appears on page 228.) Using this notation we can dispense with the special symbols for vector addition and scalar multiplication without fear of ambiguity: for example, $(a+b)(\alpha+c\beta)$ can only mean

$$(a + b) \odot [\alpha \oplus (c \odot \beta)].$$

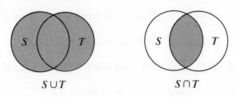

$S \cup T$ $\qquad$ $S \cap T$

FIGURE 1.6

As is customary, sets will be denoted by capital Latin letters, and the symbol $\in$ will denote the membership relation:

$x \in S$ means that the object $x$ is a member of the set $S$;
$x \notin S$ means that the object $x$ is not a member of $S$.

Sets can be described by enclosing within braces a complete list of members or a complete characterization of the elements of the set. For the latter it is often convenient to write two statements, separated by a vertical line within braces:

$$\{x \mid P(x)\}.$$

The statement before the vertical line indicates the type of elements being considered, and the statement after the vertical line specifies a property that characterizes elements of that set. For example,

$$S = \{(x, y) \in R_2 \mid x < 0 \text{ and } y > 0\}$$

specifies the set of all points of the Cartesian plane that lie in the second quadrant.

If $S$ and $T$ are sets such that each member of $S$ is a member of $T$ we write

$$S \subseteq T,$$

which states that $S$ is a *subset* of $T$. We write $S = T$ if and only if $S \subseteq T$ and $T \subseteq S$: equal sets have precisely the same members. If $S \neq T$ and $S \subseteq T$, we write $S \subset T$, which says that $S$ is a *proper* subset of $T$. The *union* of two sets is defined by

$$S \cup T = \{x \mid x \in S \text{ or } x \in T\}.$$

Thus each member of $S \cup T$ is a member of S or of $T$ or of both. The *intersection* of two sets is defined by

$$S \cap T = \{x \mid x \in S \text{ and } x \in T\}.$$

The *empty* or *void* set $\Phi$ has no members whatsoever: $\Phi$ is a subset of every set. Sets $S$ and $T$ are said to be *disjoint* if and only if $S \cap T = \Phi$.

The *Cartesian product* $S \times T$ of two sets $S$ and $T$ is the set of all ordered pairs that can be formed by choosing an element of $S$ as the first entry and an element of $T$ as the second entry:

$$S \times T = \{(x, y) \mid x \in S, y \in T\}.$$

More generally, the Cartesian product $S_1 \times S_2 \times \ldots \times S_n$ of $n$ sets $S_1, S_2, \ldots, S_n$ is the set of all ordered $n$-tuples that can be formed by choosing an element of $S_1$ as the first entry, an element of $S_2$ as the second entry, and so on. For example, the Cartesian plane is $R \times R$, and Cartesian three-space is $R \times R \times R$.

EXERCISES 1.2

1. Verify Property (5) for $R_3$.

2. Prove that there is one and only one zero vector in $R_3$.

3. Prove Properties (9–11) directly from (1–8), *without* assuming that $\alpha$ is an ordered triple of numbers. (For example, to prove Property (9), start with the special case of Property (5) in which $a = 1$ and $b = 0$; then apply successively Properties (8), (4), and (3).)

4. If $f$ and $g$ are real valued functions and if $c$ is a real number, the functions $f \oplus g$ and $c \odot f$ are defined by

$$(f \oplus g)(x) = f(x) + g(x),$$
$$(c \odot f)(x) = cf(x).$$

Regarding functions as vectors and real numbers as scalars, determine which of the following sets of functions form vector spaces:

  (i)   all real functions $f$ for which $f(1) = 0$;

  (ii)  all real functions $f$ for which $\lim\limits_{x \to 1} f(x) = 0$;

  (iii) all real functions $f$ which are continuous at $x = 1$;

  (iv)  all real functions $f$ for which $f(0) = 1$.

5. A real *polynomial* of degree not exceeding 2 is a function of the form

$$p(x) = a_0 x^2 + a_1 x + a_2,$$

where each coefficient $a_1$ is real. The sum of two polynomials and the scalar multiple of a polynomial are defined as for functions, as in Exercise 4. Show that the set $\mathscr{P}_2$ of all real polynomials of degree not exceeding 2, together with these operations, satisfies vector space properties (a), (b), and (c).

What formal similarities do you observe between $\mathscr{P}_2$ and the vector space $R_3$?

6. Write in the form $\{x \mid P(x)\}$ a description of each of the following sets:

(i)   the set $A$ of all real numbers that are integral multiples of $\pi$;

(ii)   the set $B$ of all points of $R_2$ that lie inside the circle of radius 1 with center at (2, 3);

(iii)   the set of all points of $R_3$ that lie directly above the origin.

7. Write a verbal description of each of the following sets:

(i)   $A = \{\alpha \in R_3 \mid \alpha \neq \theta\}$;

(ii)   $B = \{(x, y) \in R \times R \mid y < x\}$;

(iii)   $C = \{x \in R \mid x^2 + 1 = 0\}$.

8. Determine the number of subsets of an $n$-element set. Begin with $n = 1$, 2, 3 and proceed to prove your answer by an inductive argument.

9. Given $A = \{1, 2, 5, 7, 8\}$ and $B = \{2, 4, 5, 9\}$.

(i)   Which of the following pairs are elements of $A \times B$: (2, 9), (4, 5), (1, 8), (2, 2), (1, 1)?

(ii)   How many elements are in $A \times B$?

(iii)   How many elements are in $B \times A$?

(iv)   How many elements are in $(A \times B) \cap (B \times A)$?

10. Prove that for any sets $A$ and $B$, $A \subseteq B$ if and only if $A \cup B = B$.

11. Prove that for any sets $A$, $B$, and $C$,

$$(A \cup B) \cap C \subseteq A \cup (B \cap C);$$

moreover if $A \subseteq C$, then the subset relation can be strengthened to equality.

## 1.3   Linear Combinations of Vectors in $R_3$

Let $S$ denote any nonempty set of vectors in $R_3$. Any finite sum of scalar multiples of vectors of $S$ is called a *linear combination* of vectors of $S$. The set of *all* linear combinations of vectors of $S$ is called the set *spanned* by $S$ and is denoted $[S]$. Thus

$$[S] = \{a_1\alpha_1 + \ldots + a_p\alpha_p \mid a_i \in R, \alpha_i \in S; p = 1, 2, \ldots \}.$$

We shall now investigate the nature of the set $[S]$ for all possible choices of $S$, beginning with small sets.

If $S = \{\theta\}$, the single-element set consisting of $\theta$ alone, then $[S] = S$, since $a\theta = \theta$ for every scalar $a$, and since $\theta + \theta = \theta$. Geometrically, $[S]$ is a single point of $R_3$, namely the origin. If $S = \{\alpha\}$, a single-element set consisting of a nonzero vector $\alpha$, then $[S] = \{k\alpha \mid k \in R\}$. Geometrically, $[S]$ is a line of $R_3$, passing through the origin as shown in Figure 1.5.

If $S = \{\alpha, \beta\}$, where $\alpha \neq \theta$, either $\beta \in [\alpha]$ or $\beta \notin [\alpha]$. If $\beta \in [\alpha]$, then $\beta = k\alpha$ for some $k \in R$; so $[S] = [\alpha]$, a line through the origin. If $\beta \notin [\alpha]$, then $[S] = \{a\alpha + b\beta \mid a, b \in R\}$. Here the geometric picture is more interesting. Vectors $\alpha$ and $\beta$ determine points $A$ and $B$, which do not lie on the same line through $O$. Hence the three points $O, A$, and $B$ determine a unique plane that passes through the origin, and that plane is the geometric representation of $[S]$; every linear combination of $\alpha$ and $\beta$ corresponds to a point of that plane and, conversely, every point of the plane corresponds to some linear combination of $\alpha$ and $\beta$. (See Figure 1.4.)

If $S = \{\alpha, \beta, \gamma\}$, where $\alpha \neq \theta$ and $\beta \notin [\alpha]$, either $\gamma \in [\alpha, \beta]$ or $\gamma \notin [\alpha, \beta]$. In the former case $[\alpha, \beta]$ is a plane through the origin and $[S]$ is that plane. In the latter case, none of the three vectors $\alpha$, $\beta$, and $\gamma$ lies in the plane determined by the other two. Roughly speaking, the three geometric vectors protrude from the origin in three mutually independent directions; any vector $\xi$ of $R_3$ is a linear combination of $\alpha$, $\beta$, and $\gamma$; that is: $[S] = R_3$. See Figure 1.7.

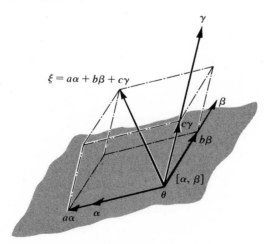

FIGURE 1.7

Thus, by forming all linear combinations of various sets of vectors in $R_3$, we obtain four types of sets of vectors:

the origin,
a line through the origin,
a plane through the origin,
the entire space $R_3$.

We see that these are the only possibilities if we let $S$ be an arbitrary non-empty set of vectors of $R_3$. If $S$ contains no nonzero vector, then $[S]$ is the origin. If $S$ contains a nonzero vector $\alpha$, then $[\alpha] \subseteq [S]$; if equality holds, $[S]$ is a line. Otherwise, there exists a vector $\beta$ in $S$ such that $\beta \notin [\alpha]$, and then $[\alpha, \beta] \subseteq [S]$. Again, if equality holds, $[S]$ is a plane, but otherwise there exists a vector $\gamma$ in $S$ such that $\gamma \notin [\alpha, \beta]$. Hence $[\alpha, \beta, \gamma] \subseteq [S]$. But then equality must hold since $[S] \subseteq R_3 = [\alpha, \beta, \gamma]$. This equality is based on geometric intuition; a proof will be given later.

It follows that $R_3$ is spanned by a properly chosen set of *three* vectors and that no set of fewer than three vectors will span $R_3$. The number of vectors in a smallest spanning set is used later to define the *dimension* of any finite-dimensional vector space, and at that time we shall prove the intuitive assertions of this section.

An obvious example of a spanning set for $R_3$ consists of the vectors

$$\epsilon_1 = (1, 0, 0),$$

$$\epsilon_2 = (0, 1, 0),$$

$$\epsilon_3 = (0, 0, 1).$$

Clearly $(a_1, a_2, a_3) = a_1 \epsilon_1 + a_2 \epsilon_2 + a_3 \epsilon_3$, so the set $\{\epsilon_1, \epsilon_2, \epsilon_3\}$ spans $R_3$.

Another example of a spanning set is given by

$$\beta_1 = (1, -1, 0),$$

$$\beta_2 = (0, 1, -2),$$

$$\beta_3 = (1, 0, -1),$$

but in this set it is not quite so easy to discover how to represent an arbitrary vector as a linear combination of $\beta_1$, $\beta_2$, and $\beta_3$. We must determine scalars $c_1$, $c_2$, and $c_3$ such that

$$(a_1, a_2, a_3) = c_1\beta_1 + c_2\beta_2 + c_3\beta_3$$
$$= c_1(1, -1, 0) + c_2(0, 1, -2) + c_3(1, 0, -1)$$
$$= (c_1 + c_3, -c_1 + c_2, -2c_2 - c_3).$$

Hence we wish to solve the following system of linear equations:

$$c_1 \qquad + c_3 = a_1,$$
$$-c_1 + \ c_2 \qquad = a_2,$$
$$-2c_2 - c_3 = a_3.$$

The unique solution is easily determined to be

$$c_1 = -a_1 - 2a_2 - a_3,$$
$$c_2 = -a_1 - \ a_2 - a_3,$$
$$c_3 = 2a_1 + 2a_2 + a_3.$$

On the other hand, not every set of three vectors in $R_3$ will be a spanning set. For example, if $\beta_3$ is replaced by $\gamma_3 = (4, -7, 6)$, the set $\{\beta_1, \beta_2, \gamma_3\}$ is not a spanning set because $\gamma_3 = 4\beta_1 - 3\beta_2$. Geometrically, $\gamma_3$ lies in the plane spanned by $\beta_1$ and $\beta_2$.

Consider again the four types of sets of vectors that can be obtained in $R_3$ by forming all linear combinations of various nonempty sets of vectors; $[S]$ is either a single point $\theta$ (the origin), a line through $\theta$, a plane through $\theta$, or $R_3$ itself. If we reexamine the list of Properties (1–8) of Section 1.2 valid for any vector space, we see precisely how the origin differs from any other point of the space and how lines and planes through the origin are distinguished from other lines and planes. These distinctions become evident if we ask the question: Under what conditions does a subset $S$ of $R_3$ form a vector space relative to the algebraic operations defined for $R_3$? To begin with, Property (3) asserts that $S$ must contain at least one vector, $\theta$. Moreover, a vector space must contain all linear combinations of its vectors, which means that for a nonempty set $S$ of vectors to form a vector space it is necessary that $S = [S]$. This condition is sufficient, too, because if $S = [S]$, then for any $\alpha \in S$, $0\,\alpha = \theta$ and $-1\,\alpha = -\alpha \in S$. Thus, Properties (3) and (4) are satisfied. Properties (1–2) and (5–8) are universal properties, which hold for all vectors in $R_3$ and all scalars in $R$, and which therefore hold for all vectors in $S$ and for all scalars in $R$.

The term *subspace* is used to designate a subset of a vector space that itself forms a vector space relative to the operations of vector sum and scalar multiple as defined for the parent space. Our discussion has established the following two statements about subspaces of $R_3$.

(a)  A subset $S \subseteq R_3$ is a subspace of $R_3$ if and only if $S = [S]$.

(b)  Any subspace of $R_3$ is either
the zero space $\{\theta\}$,
a line $\{k \, \alpha \mid \alpha \neq \theta, k \in R\}$ through the origin,
a plane $\{a \, \alpha + b \, \beta \mid \alpha \neq \theta, \beta \notin [\alpha]; a, b \in R\}$ through the origin,
or $R_3$ itself.

It is worth remarking that the arguments used to establish (a) made use of properties that are true in any vector space; never did the specific nature of ordered triples enter our consideration. Hence (a) is a theorem that is valid in any vector space. Also $[S]$ can now be called the *subspace spanned by S.*

It should be intuitively clear that each line through the origin in $R_3$ can be regarded as the vector space $R$ and that each plane through the origin can be regarded as the vector space $R_2$. A study of $R_3$ therefore includes a study of $R$ and $R_2$; if we prefer to select specific examples of a line and a plane, both of which pass through the origin, we can conveniently use the subspaces

$$\{(a_1, 0, 0) \mid a_1 \in R\}$$

and

$$\{(a_1, a_2, 0) \mid a_1, a_2 \in R\}.$$

## EXERCISES 1.3

1. Given the four vectors $\alpha, \beta, \gamma$, and $\delta$, describe geometrically each of the following subspaces: $[\alpha, \beta], [\gamma, \delta], [\alpha, \beta] \cap [\gamma, \delta], [\alpha, \beta] \cap [\delta]$.

(i)   $\alpha = (1, 1, 0)$    $\beta = (0, 0, 1)$    $\gamma = (1, 0, 0)$    $\delta = (1, 2, 1)$.

(ii)  $\alpha = (2, -1, 1)$    $\beta = (1, 1, 3)$    $\gamma = (1, -2, -2)$    $\delta = (5, -4, 0)$.

(iii) $\alpha = (1, -1, 1)$    $\beta = (3, 4, -2)$    $\gamma = (4, 3, -1)$    $\delta = (1, -8, 6)$.

2. Given the vectors $\alpha, \beta$, and $\gamma$, describe geometrically $[\alpha, \beta, \gamma]$.

(i)  $\alpha = (1, -1, 2)$    $\beta = (1, 0, -1)$    $\gamma = (0, 1, 11)$.

(ii) $\alpha = (1, 1, -8)$    $\beta = (3, -1, -5)$    $\gamma = (1, -3, 11)$.

3. Which of the following sets are subspaces of $R_3$? Give a reason for each answer.

(i) $\{(a_1, a_2, a_3) \mid a_1 = 0\}$.

(ii) $\{(a_1, a_2, a_3) \mid a_1 = 1\}$.

(iii) $\{(a_1, a_2, a_3) \mid a_3 = 2a_1 - a_2\}$.

(iv) $\{(a_1, a_2, a_3) \mid a_3 = a_2{}^2\}$.

(v) $\{(a_1, a_2, a_3) \mid a_1 + a_2 + a_3 = 0\}$.

(vi) $\{(a_1, a_2, a_3) \mid a_1 + a_2 + a_3 = 1\}$.

4. For each case in which it is possible, express $\xi = (1, 1, -12)$ as a linear combination of the vectors $\alpha, \beta, \gamma$ specified in

(i) Exercise 1(i),

(ii) Exercise 2(i),

(iii) Exercise 2(ii).

5. Let $\alpha$ and $\beta$ be as defined in Exercise 1(iii). Write an equation for the plane $[\alpha, \beta]$. (*Hint:* A point $(x_1, x_2, x_3)$ lies on that plane if and only if there are scalars $a$ and $b$ such that $(x_1, x_2, x_3) = a\,\alpha + b\,\beta$. This yields three equations in $a$ and $b$. By eliminating $a$ and $b$ you can obtain a linear equation in $x_1, x_2$, and $x_3$.)

6. Given $\alpha = (1, -1, 1)$, $\beta = (3, 4, -2)$, $\gamma = (9, 5, -1)$, and $\delta = (-17, -11, 3)$. Show that any vector that is a linear combination of $\alpha$ and $\beta$ is also a linear combination of $\gamma$ and $\delta$ and conversely.

7. Prove that a nonvoid subset $S$ of $R_3$ is a subspace if and only if for every $c \in R$ and all $\alpha, \beta \in S$, $c\,\alpha \in S$ and $\alpha + \beta \in S$.

8. Describe two different spanning sets for the space of all real polynomials of degree not exceeding 2, described in Exercise 5 of Section 1.2.

9. If $\mathscr{S}$ and $\mathscr{T}$ are subspaces of $R_3$, is $\mathscr{S} \cap \mathscr{T}$ a subspace? Is $\mathscr{S} \cup \mathscr{T}$ a subspace? Prove each of your answers.

10. Prove: Let $\mathscr{S}$ and $\mathscr{T}$ be subspaces of $R_3$. If $\mathscr{S} \cup \mathscr{T}$ (set union) is also a subspace of $R_3$, then either $\mathscr{S} \subseteq \mathscr{T}$ or $\mathscr{T} \subseteq \mathscr{S}$. (The converse is obviously valid.)

# 1.4   Systems of Linear Equations

In the previous section we have seen that an investigation of subspaces and spanning sets in $R_3$ leads naturally to the problem of solving a system of several linear equations in three unknowns. Given a finite set $S$ of vectors

$$\alpha_1 = (a_{11}, a_{21}, a_{31}),$$

$$\alpha_2 = (a_{12}, a_{22}, a_{32}),$$

.

.

.

$$\alpha_n = (a_{1n}, a_{2n}, a_{3n}),$$

a vector $\delta = (d_1, d_2, d_3)$ is in the subspace $[S]$ spanned by $S$ if and only if $\delta$ is a linear combination of vectors of $S$. That is, $\delta \in [S]$ if and only if there exist scalars $x_1, x_2, \ldots, x_n$ such that

$$\delta = x_1\alpha_1 + x_2\alpha_2 + \ldots + x_n\alpha_n = \sum_{i=1}^{n} x_i\alpha_i.$$

Written in terms of components, this linear vector equation becomes a system of three linear scalar equations in $n$ unknowns:

$$a_{11}x_1 + a_{12}x_2 + \ldots + a_{1n}x_n = d_1,$$

(1.1)
$$a_{21}x_1 + a_{22}x_2 + \ldots + a_{2n}x_n = d_2,$$

$$a_{31}x_1 + a_{32}x_2 + \ldots + a_{3n}x_n = d_3.$$

If $(d_1, d_2, d_3)$ is given, then $\delta \in [S]$ only if there exists an $n$-tuple of scalars $(x_1, \ldots, x_n)$ satisfying this system of equations. Conversely, if $(x_1, \ldots, x_n)$ is any $n$-tuple of scalars, then the vector $(d_1, d_2, d_3)$ whose components are defined by this system of equations must be in $[S]$.

The need to solve such systems arises so frequently in linear algebra that we need to find an efficient procedure of computation that guarantees a solution. Any method of computation that solves a specific type of problem in a finite number of steps is called an *algorithm*. Although many algorithms exist for solving linear systems, the most efficient method turns out to be a procedure for successive elimination of variables, introduced by Gauss. The idea of *Gaussian elimination* is to solve one of the equations for $x_1$ in terms of the other $x$'s and to substitute that expression for $x_1$ in all of the other equations, thereby obtaining a new system with the same solutions as the original and with the form

$$b_{11}x_1 + b_{12}x_2 + \ldots + b_{1n}x_n = e_1,$$

$$b_{22}x_2 + \ldots + b_{2n}x_n = e_2,$$

$$b_{32}x_2 + \ldots + b_{3n}x_n = e_3,$$

where the $b$'s and $e$'s are known combinations of the $a$'s and $d$'s. The process is then repeated, using the second equation to solve for another variable, say $x_2$, and substituting the result into the third equation to produce an equivalent system in the *echelon* form

$$c_{11}x_1 + c_{12}x_2 + c_{13}x_3 + \ldots + c_{1n}x_n = f_1,$$

$$c_{22}x_2 + c_{23}x_3 + \ldots + c_{2n}x_n = f_2,$$

$$c_{33}x_3 + \ldots + c_{3n}x_n = f_3.$$

Once the system is brought into echelon form, the solution can be completed easily by the process of *back substitution*: the last equation can be solved for $x_3$ in terms of $x_4, \ldots, x_n$ and the known coefficients; the second equation can be solved for $x_2$ in terms of $x_4, \ldots, x_n$ and known coefficients, using the expression already determined for $x_3$; finally, the first equation can be solved for $x_1$ in terms of $x_4, \ldots, x_n$ and known coefficients, using the expressions already determined for $x_3$ and $x_2$.

Fortunately, the application of this algorithm is even simpler than its general description, and further shortcuts in computation can be achieved by using matrix notation. For instance, observe first the pattern of coefficients in the original System (1.1): the components of $\alpha_1$ occur in a vertical column as coefficients of $x_1$, the components of $\alpha_2$ occur in the next column as coefficients of $x_2$, and similarly for $\alpha_3, \ldots, \alpha_n$. Hence it is convenient to represent each vector in *column* form:

$$\alpha_1 = \begin{pmatrix} a_{11} \\ a_{21} \\ a_{31} \end{pmatrix}, \; \alpha_2 = \begin{pmatrix} a_{12} \\ a_{22} \\ a_{32} \end{pmatrix}, \; \ldots, \; \alpha_n = \begin{pmatrix} a_{1n} \\ a_{2n} \\ a_{3n} \end{pmatrix}, \; \delta = \begin{pmatrix} d_1 \\ d_2 \\ d_3 \end{pmatrix}.$$

Next, observe that the operations of Gaussian elimination are completely described by these scalars; that is, the $x$'s need not be written since they serve only to number the various columns during the computation. Hence all the essential information of the linear system can be represented by an array of scalars, or a *matrix*, of the form

$$\begin{pmatrix} a_{11} \; a_{12} \; \ldots \; a_{1n} & d_1 \\ a_{21} \; a_{22} \; \ldots \; a_{2n} & d_2 \\ a_{31} \; a_{32} \; \ldots \; a_{3n} & d_3 \end{pmatrix}.$$

To apply the method of Gaussian elimination, if $a_{11} \neq 0$ we solve the first equation of System (1.1) for $x_1$ and substitute this expression for $x_1$ in the subsequent equations, obtaining System (1.2):

(1.2)
$$
\begin{aligned}
x_1 + \quad & a_{11}^{-1}a_{12}x_2 + \ldots + && a_{11}^{-1}a_{1n}x_n = a_{11}^{-1}d_1, \\
& (a_{22} - a_{21}a_{11}^{-1}a_{12})x_2 + \ldots + (a_{2n} - a_{21}a_{11}^{-1}a_{1n})x_n = d_2 - a_{21}a_{11}^{-1}d_1, \\
& (a_{32} - a_{31}a_{11}^{-1}a_{12})x_2 + \ldots + (a_{3n} - a_{31}a_{11}^{-1}a_{1n})x_n = d_3 - a_{31}a_{11}^{-1}d_1.
\end{aligned}
$$

If each equation is multiplied by $a_{11}$, the resulting matrix becomes

$$
\begin{pmatrix}
a_{11} & a_{12} & \ldots & a_{1n} & \vdots & d_1 \\
0 & b_{22} & \ldots & b_{2n} & \vdots & e_2 \\
0 & b_{32} & \ldots & b_{3n} & \vdots & e_3
\end{pmatrix},
$$

where $b_{ij} = a_{11}a_{ij} - a_{i1}a_{1j}$ and $e_i = a_{11}d_i - a_{i1}d_1$ for $i, j \geq 2$.

If $b_{22} \neq 0$, we can apply the elimination process to the system consisting of the second and third equations to obtain a matrix in echelon form

$$
\begin{pmatrix}
a_{11} & a_{12} & a_{13} & \ldots & a_{1n} & \vdots & d_1 \\
0 & b_{22} & b_{23} & \ldots & b_{2n} & \vdots & e_2 \\
0 & 0 & c_{33} & \ldots & c_{3n} & \vdots & f_3
\end{pmatrix},
$$

where $c_{3j} = b_{22}b_{3j} - b_{32}b_{2j}$ for $j \geq 3$ and $f_3 = b_{22}e_3 - b_{32}e_2$.

When the system (or its matrix) is reduced to echelon form, solutions are easily obtained: let $x_4, \ldots, x_n$ be *any* numbers; then

$$
x_3 = c_{33}^{-1}(f_3 - c_{34}x_4 - \ldots - c_{3n}x_n),
$$
$$
x_2 = b_{22}^{-1}(e_2 - b_{23}x_3 - \ldots - b_{2n}x_n),
$$
$$
x_1 = a_{11}^{-1}(d_1 - a_{12}x_2 - \ldots - a_{1n}x_n).
$$

As an example, consider the system obtained in Section 1.3 to show that the given set $\{\beta_1, \beta_2, \beta_3\}$ spans $R_3$. The matrix of the original system is

$$
\begin{pmatrix}
1^* & 0 & 1 & \vdots & a_1 \\
-1 & 1 & 0 & \vdots & a_2 \\
0 & -2 & -1 & \vdots & a_3
\end{pmatrix}.
$$

The first Gaussian step proceeds from the starred element (called a *pivot*) in the first row and first column, and by using System (1.2) the resulting matrix may be calculated to be

$$\begin{pmatrix} 1 & 0 & 1 & \vdots & a_1 \\ 0 & 1^* & 1 & \vdots & a_2 + a_1 \\ 0 & -2 & -1 & \vdots & a_3 \end{pmatrix}.$$

A second pivot, on the element in the second row and second column, yields the echelon form

$$\begin{pmatrix} 1 & 0 & 1 & \vdots & a_1 \\ 0 & 1 & 1 & \vdots & a_2 + a_1 \\ 0 & 0 & 1 & \vdots & a_3 + 2(a_2 + a_1) \end{pmatrix}.$$

Now the solution is obtained:

$$x_3 = 2a_1 + 2a_2 + a_3,$$

$$x_2 = a_1 + a_2 - x_3 = -a_1 - a_2 - a_3,$$

$$x_1 = a_1 - x_3 = -a_1 - 2a_2 - a_3,$$

as asserted in Section 1.3.

Further analysis of the Gaussian method yields another formal procedure for Gaussian reduction of a matrix to echelon form. To solve one system of linear equations, it suffices to solve any other system that has *precisely* the same set of solutions; the Gaussian algorithm describes a means of trading one system for an equivalent system in different form, continuing until we obtain a system that is easy to solve. The solutions of a system will remain unchanged by the performance of any of the following operations:

(M) Replacing any equation with the equation obtained by *multiplying* each term of the given equation by the same nonzero constant;

(A) Replacing any equation with the equation obtained by *adding* the coefficients of like terms of the given equation and any other equation of the system; and

(P) *Permuting* the order in which the equations are written.

It should be emphasized that here we are considering exact solutions of the system. Computer-generated solutions, which are approximate, can be changed drastically by these operations, and writing programs for automatic computation requires special care. (See Exercise 6.)

Consideration of our detailed calculations for a single step in Gaussian elimination makes clear that, given a nonzero pivot element, each step can be performed by operations of type (M) and (A). Operations of type (P) can be used to obtain a nonzero pivot in the preferred position. Hence, the method of Gaussian elimination can be carried out by reducing the matrix of the system to row echelon form by a sequence of *elementary row operations* of the three given types. We note that successive operations of types (M) and (A) can be used to replace any equation with a linear combination of the equations of the system, provided the given equation has a nonzero coefficient in that linear combination.

As an example, we solve the system

$$x_2 + x_3 = 5,$$

$$2x_1 + x_2 + 5x_3 = 1,$$

$$x_1 + x_2 + 3x_3 = 3.$$

Although there is no compelling reason to begin with a pivot element in the first row and first column, we can arrange to do so by interchanging the positions of the first and third equations to obtain the matrix

$$\begin{pmatrix} 1 & 1 & 3 & \vdots & 3 \\ 2 & 1 & 5 & \vdots & 1 \\ 0 & 1 & 1 & \vdots & 5 \end{pmatrix}.$$

Denote row $i$ by $R_i$. Replacing $R_2$ with $R_2 - 2R_1$, we have

$$\begin{pmatrix} 1 & 1 & 3 & \vdots & 3 \\ 0 & -1 & -1 & \vdots & -5 \\ 0 & 1 & 1 & \vdots & 5 \end{pmatrix}.$$

Replace $R_3$ with $R_3 + R_2$ to obtain a matrix in echelon form:

$$\begin{pmatrix} 1 & 1 & 3 & \vdots & 3 \\ 0 & -1 & -1 & \vdots & -5 \\ 0 & 0 & 0 & \vdots & 0 \end{pmatrix}.$$

Any values of $x_1$, $x_2$, and $x_3$ satisfy the equation represented by the third row, so we obtain as the solution:

$$x_3 \text{ arbitrary},$$

$$x_2 = 5 - x_3,$$

$$x_1 = 3 - 3x_3 - x_2 = -2 - 2x_3.$$

Finally, we observe that the system in this example is easily solved directly without the use of Gaussian elimination or matrix notation. The value of having an effective algorithm is more evident when one considers a large system of perhaps 100 linear equations in 120 unknowns.

We shall return to the problem of solving systems of linear equations in Chapter 4, which contains a treatment more theoretical than the one appearing here.

## EXERCISES 1.4

1. Solve the following systems of linear equations by the method of Gaussian elimination.

$$
\text{(i)} \quad
\begin{aligned}
x_1 - x_2 + 2x_3 &= 3, \\
3x_1 - 4x_2 + 5x_3 &= 9, \\
x_1 + x_2 + x_3 &= 6.
\end{aligned}
$$

$$
\text{(ii)} \quad
\begin{aligned}
x_1 + 2x_2 + 4x_3 &= 7, \\
x_1 \qquad\;\; + 2x_3 &= -2, \\
2x_1 + 3x_2 + 7x_3 &= 9.
\end{aligned}
$$

$$
\text{(iii)} \quad
\begin{aligned}
x_1 - x_2 \qquad\quad &= 2, \\
-x_1 + x_2 + 2x_3 &= -1, \\
x_1 - x_2 + 4x_3 &= 4.
\end{aligned}
$$

2. Determine the subspace of $R_3$ that is spanned by the vectors

$$\alpha_1 = (1, 1, 2),$$
$$\alpha_2 = (2, 0, 3),$$
$$\alpha_3 = (4, 2, 7).$$

Does the vector $(7, -2, 9)$ belong to that subspace? Relate your conclusions to the results of Exercise 1(ii).

3. Suppose that in reducing the matrix of the linear system (1.1) to echelon form you obtain a row in which the only nonzero entry appears in the last column. What does that mean about the solutions of the linear system? Explain.

4. Suppose that in reducing the matrix of the linear system (1.1) to echelon form you obtain a row in which every entry is zero. What does that mean about the solutions of the linear system? Explain.

5. Let $A$ and $B$ be three-by-three real matrices. Show that if each row of $B$ is a linear combination of the rows of $A$, then $B$ can be obtained from $A$ by a finite sequence of elementary row operations. Is the converse true? Explain.

6. For an account of some of the hazards in computing read one or more of the following articles:

Forsythe, G. E., "Solving a Quadratic Equation on a Computer." In *The Mathematical Sciences, A Collection of Essays*. Committee on Support of Research in the Mathematical Sciences. Cambridge, Mass.: M.I.T. Press (1969). Pages 138–152.

Forsythe, G. E., "Pitfalls in Computation, or Why a Math Book Isn't Enough." *The American Mathematical Monthly*, 77, 931–956 (1970).

Conte, S. D. *Elementary Numerical Analysis*. New York: McGraw-Hill (1965). Chapter 1.

## 1.5   Linear Independence and Bases in $R_3$

In Section 1.3 our attention centered on the problem of determining the subspace $[S]$ spanned by a given nonempty subset $S$ of $R_3$. A natural extension of this problem is to find a smallest subset $T$ of $S$ such that $T$ and $S$ span the same subspace.

If $S$ contains a nonzero vector $\alpha_1$, let $T_1 = \{\alpha_1\}$; observe that $[T_1] \subseteq [S]$. If $\xi \in [T_1]$ for every $\xi \in S$, then $[T_1] = [S]$, and the job is done. Otherwise, there exists $\alpha_2 \in S$ such that $\alpha_2 \notin [T_1]$. Let $T_2 = \{\alpha_1, \alpha_2\}$. Again $[T_2] \subseteq [S]$; if $\xi \in [T_2]$ for every $\xi \in S$, then $[T_2] = [S]$. Otherwise there exists $\alpha_3 \in S$ such that $\alpha_3 \notin [T_2]$. Let $T_3 = \{\alpha_1, \alpha_2, \alpha_3\}$. Our analysis of the subspaces of $R_3$ assures us that no larger set is needed, since $[T_3] = R_3$.

Note that at each stage of the foregoing construction we chose $\alpha_{k+1} \notin [\alpha_1, \ldots, \alpha_k]$ whenever such a vector existed. This guaranteed that no chosen vector was a linear combination of vectors previously chosen. Thus we made sure that, of all linear equations of the form

$$c_1\alpha_1 + \ldots + c_p\alpha_p = \theta,$$

none could hold for the chosen vectors except the trivial equation in which each $c_i = 0$. To confirm this assertion, suppose there exists a set of scalars $c_i$, *not all* of which are 0, such that

$$c_1\alpha_1 + \ldots + c_p\alpha_p = \theta.$$

Let $c_{k+1}$ be the last nonzero coefficient in that equation. Then

$$\alpha_{k+1} = -c_{k+1}^{-1}(c_1\alpha_1 + \ldots + c_k\alpha_k),$$

contrary to the selection of $\alpha_{k+1} \notin [\alpha_1, \ldots, \alpha_k]$. Hence $c_i = 0$ for

$i = 1, \ldots, p$. This concept is of basic importance in linear algebra, and we adopt special terminology for it.

A set $S$ of vectors is said to be *linearly dependent* if and only if there exists a finite set $\{\alpha_1, \ldots, \alpha_k\}$ of vectors of $S$ and a finite set of scalars $\{c_1, \ldots, c_k\}$ *not all of which are zero* such that

$$c_1\alpha_1 + c_2\alpha_2 + \ldots + c_k\alpha_k = \theta.$$

Otherwise, $S$ is said to be *linearly independent*.

Much of our future work makes essential use of these definitions, so you should be sure that you understand them thoroughly. A set of vectors is linearly dependent if and only if *some* nontrivial linear combination of those vectors is the zero vector. On the other hand, a set of vectors is linearly independent if and only if the zero vector can be constructed as a linear combination of the set *only* by selecting *every* coefficient of the linear combination to be 0. To prove that a set $\{\alpha_1, \ldots, \alpha_m\}$ is linearly independent we shall frequently assume that scalars $c_1, \ldots, c_m$ exist such that

$$c_1\alpha_1 + \ldots + c_m\alpha_m = \theta,$$

and then deduce that each $c_i$ must be zero.

Our knowledge of $R_3$ easily yields a geometric description of all linearly independent subsets of $R_3$. Any one-element set except $\{\theta\}$ is linearly independent. A two-element set $\{\alpha, \beta\}$ is linearly independent if and only if $\alpha$ and $\beta$ do not lie on one line through the origin. A three-element set $\{\alpha, \beta, \gamma\}$ is linearly independent if and only if $\alpha$, $\beta$, and $\gamma$ do not lie on one plane through the origin. The only other linearly independent subset of $R_3$ is the empty set $\Phi$. Although the claim that $\Phi$ is linearly independent might at first be contrary to our intuition, it is easy to verify that $\Phi$ does satisfy the formal definition of linear independence because it does *not* satisfy the definition of linear dependence.

A linearly independent set of one vector spans a line through the origin. A linearly independent set of two vectors spans a plane through the origin. A linearly independent set of three vectors spans $R_3$. The only other subspace of $R_3$ is $\{\theta\}$, and the only other linearly independent subset of $R_3$ is $\Phi$. It is convenient therefore to extend our previous definition of the subspace $[S]$ spanned by a set $S$ to allow $S$ to be the empty set and arbitrarily to define $[\Phi] = \{\theta\}$. Then every subspace of $R_3$ is spanned by at least one linearly independent subset of $R_3$.

A *basis* for a subspace $\mathscr{S}$ of $R_3$ is a linearly independent set of vectors that spans $\mathscr{S}$.

Thus a set $B$ of vectors of $R_3$ is a basis for a subspace $\mathscr{S}$ of $R_3$ if and only if $B$ satisfies two conditions:

  (a)  $B$ is linearly independent, and

  (b)  $[B] = \mathscr{S}$.

In a geometric coordinate system it is necessary to know not only that certain directed lines are designated as coordinate axes, but also that a sequential order is assigned to these axes. Thus, to relate the algebraic concept of a basis to the geometric concept of a coordinate system it is necessary to write the elements of a basis in some order that remains fixed throughout the discussion. A basis in which a sequential order is assigned to the basis vectors is called an *ordered* basis. When we use the term *basis* in this book, we shall mean *ordered basis*. Hence, we shall regard the linearly independent sets $\{\alpha_1, \alpha_2, \alpha_3\}$ and $\{\alpha_3, \alpha_2, \alpha_1\}$ as different bases for $R_3$ even though they are equal as subsets of $R_3$.

### Examples of Ordered Bases for $R_3$.

$$\begin{cases} \epsilon_1 = (1, 0, 0), \\ \epsilon_2 = (0, 1, 0), \\ \epsilon_3 = (0, 0, 1). \end{cases} \quad \begin{cases} \alpha_1 = (0, 1, 1), \\ \alpha_2 = (1, 0, 1), \\ \alpha_3 = (1, 1, 0). \end{cases} \quad \begin{cases} \beta_1 = (1, 0, 0), \\ \beta_2 = (1, 1, 0), \\ \beta_3 = (1, 1, 1). \end{cases}$$

In order to verify that any of the given sets is a basis we must show that the set is linearly independent and spans $R_3$. Later we shall prove a theorem that shows that a set of *three* vectors of $R_3$ is linearly independent if and only if it spans $R_3$. But to see the nature of the calculations involved in each of these proofs, we now show that $\{\beta_1, \beta_2, \beta_3\}$ is linearly independent and also that it spans $R_3$. To prove linear independence we suppose that scalars $b_1$, $b_2$, $b_3$ exist such that

$$b_1\beta_1 + b_2\beta_2 + b_3\beta_3 = \theta.$$

In terms of the components of the given vectors this yields a system of three linear equations in the three scalars $b_1$, $b_2$, and $b_3$:

$$b_1 + b_2 + b_3 = 0,$$

$$b_2 + b_3 = 0,$$

$$b_3 = 0.$$

The only solution is $b_3 = b_2 = b_1 = 0$.

To prove that $\{\beta_1, \beta_2, \beta_3\}$ spans $R_3$ we must show that any vector $\gamma = (c_1, c_2, c_3)$ is a linear combination of the $\beta_i$: for suitable $b_i$

$$b_1\beta_1 + b_2\beta_2 + b_3\beta_3 = \gamma.$$

In terms of components this yields a system of linear equations in the $b_i$, closely related to the system obtained above:

$$b_1 + b_2 + b_3 = c_1,$$
$$b_2 + b_3 = c_2,$$
$$b_3 = c_3.$$

For any numbers $c_1$, $c_2$, and $c_3$, there is a unique solution, namely

$$b_3 = c_3,$$
$$b_2 = c_2 - c_3,$$
$$b_1 = c_1 - c_2.$$

The $\epsilon$-basis is called the *standard basis* or the *natural basis* for $R_3$. In order to relate the concept of a numerical vector in $R_3$ (an ordered triple of scalars) to the concept of a geometric vector (a directed line segment starting at the origin) we must select a coordinate system for three-space; for convenience we frequently choose the standard system, in which the three axes are mutually perpendicular. This coordinate system corresponds to the standard basis, and the point $(a, b, c)$ in that coordinate system corresponds to the vector $a\epsilon_1 + b\epsilon_2 + c\epsilon_3$. Specifically, we have

$$\alpha_1 = \quad\quad \epsilon_2 + \epsilon_3,$$
$$\alpha_2 = \epsilon_1 \quad\quad + \epsilon_3,$$
$$\alpha_3 = \epsilon_1 + \epsilon_2 \quad .$$

As you can demonstrate in Exercise 1, any vector can be expressed uniquely as a linear combination of the vectors of any basis, so we can express the $\alpha_j$ also in terms of the $\beta_i$. To do so we must determine nine scalars $c_{ij}$ such that

$$\alpha_j = c_{1j}\beta_1 + c_{2j}\beta_2 + c_{3j}\beta_3, \quad j = 1, 2, 3.$$

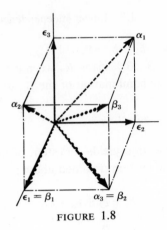

FIGURE 1.8

This is a system of three vector equations

$$(0, 1, 1) = c_{11}(1, 0, 0) + c_{21}(1, 1, 0) + c_{31}(1, 1, 1),$$
$$(1, 0, 1) = c_{12}(1, 0, 0) + c_{22}(1, 1, 0) + c_{32}(1, 1, 1),$$
$$(1, 1, 0) = c_{13}(1, 0, 0) + c_{23}(1, 1, 0) + c_{33}(1, 1, 1).$$

The first of these equations yields a system of three linear equations

$$0 = 1c_{11} + 1c_{21} + 1c_{31},$$
$$1 = 0c_{11} + 1c_{21} + 1c_{31},$$
$$1 = 0c_{11} + 0c_{21} + 1c_{31},$$

for which the only solution is $c_{31} = 1$, $c_{21} = 0$, $c_{11} = -1$. The other two vector equations can be solved in the same way to obtain

$$\alpha_1 = -1\beta_1 + 0\beta_2 + 1\beta_3,$$
$$\alpha_2 = \phantom{-}1\beta_1 - 1\beta_2 + 1\beta_3,$$
$$\alpha_3 = \phantom{-}0\beta_1 + 1\beta_2 + 0\beta_3.$$

It is instructive to verify these results geometrically as in Figure 1.8. Thus we see that the vector $\alpha_1$ can be expressed as

$$\alpha_1 = 0\epsilon_1 + 1\epsilon_2 + 1\epsilon_3$$
$$= 1\alpha_1 + 0\alpha_2 + 0\alpha_3$$
$$= -1\beta_1 + 0\beta_2 + 1\beta_3.$$

In this sense the three number triples (0, 1, 1), (1, 0, 0), and (−1, 0, 1) all serve to represent the same geometric vector relative to different coordinate systems. In order to reduce this ambiguity we can use square brackets to denote coordinates of a vector relative to some basis other than the standard basis, reserving parentheses to signify the standard basis. Thus

$$(a_1, a_2, a_3) \text{ denotes } a_1\epsilon_1 + a_2\epsilon_2 + a_3\epsilon_3;$$

$$[a_1, a_2, a_3] \text{ denotes } a_1\alpha_1 + a_2\alpha_2 + a_3\alpha_3.$$

The representation of each vector of $R_3$ as a unique linear combination of the vectors of any basis shows that the algebraic concept of a basis is the counterpart of the geometric concept of a coordinate system. Suppose we had chosen for $R_3$ the skew coordinate system whose axes are along the directions of $\beta_1, \beta_2, \beta_3$ and whose unit points on these axes are the terminal points of $\beta_1, \beta_2, \beta_3$. The above calculations show, for example, that $\alpha_1$, whose coordinates are (0, 1, 1) in the $\epsilon$-system and [1, 0, 0] in the $\alpha$-system has coordinates [−1, 0, 1] in the $\beta$-system.

## EXERCISES 1.5

1. Prove that each vector of $R_3$ can be represented in one and only one way as a linear combination of vectors of a given basis. (*Hint:* let $\{\alpha_1, \alpha_2, \alpha_3\}$ be a basis; suppose that

$$\beta = c_1\alpha_1 + c_2\alpha_2 + c_3\alpha_3 = d_1\alpha_1 + d_2\alpha_2 + d_3\alpha_3,$$

and use linear independence to deduce that $d_i = c_i$ for $i = 1, 2, 3$.)

2. Referring to the examples of bases given in this section, express each $\epsilon_j$ as a linear combination of the $\beta_i$.

3. Refer to the examples of bases given in this section. If $\xi = 5\epsilon_1 - 4\epsilon_2 + 3\epsilon_3$, express $\xi$ in terms of the $\beta$-basis. If $\eta = 5\beta_1 - 4\beta_2 + 3\beta_3$, express $\eta$ in terms of the $\epsilon$-basis.

4. Prove the following assertions:

(i) if $\theta$ is a member of a set $S$ of vectors, then $S$ is linearly dependent;

(ii) any subset of a linearly independent set must also be linearly independent;

(iii) if $S$ contains a linearly dependent subset, then $S$ is linearly dependent.

5. Prove that if $\{\alpha_1, \ldots, \alpha_p\}$ is linearly independent, and if $c_j \neq 0$ for $j = 1, \ldots, p$, then $\{c_1\alpha_1, \ldots, c_p\alpha_p\}$ is linearly independent. Interpret this result geometrically in $R_3$ for $p = 3$.

6. For each set of four vectors given in Exercise 1 of Section 1.3, select a maximal linearly independent subset and express the remaining members of the set as a linear combination of the selected vectors.

7. Determine all values of $x$ such that the set $\{(2, -3, 1), (-4, 1, 2), (0, -5, x)\}$ is linearly dependent. Explain geometrically why there is one and only one value for $x$.

8. Determine all values of $k$ such that the set $\{(1, 2, k), (0, 1, k-1), (3, 4, 3)\}$ is linearly independent.

9. Let $\gamma_1 = (1, 1, 2)$; $\gamma_2 = (2, 2, 1)$; and $\gamma_3 = (1, 2, 2)$.

(i) Show that $\{\gamma_1, \gamma_2, \gamma_3\}$ is a basis for $R_3$.

(ii) Express each vector of the standard basis as a linear combination of the $\gamma_i$.

(iii) Express $(-2, -3, 2)$ as a linear combination of the $\gamma_i$.

10. In the polynomial space $\mathscr{P}_2$ let $r$, $s$, $t$ be distinct real numbers and let $S = \{(x - r)(x - s), (x - r)(x - t), (x - s)(x - t)\}$. Is $S$ linearly independent? Show your reasoning fully.

11. If $\xi$ has coordinates $[1, -2, 3]$ relative to the $\alpha$-basis described in this section, determine the coordinates of $\xi$

(i) relative to the $\epsilon$-basis,

(ii) relative to the $\beta$-basis.

## 1.6    The Scalar Product in $R_3$

If we review carefully the notions introduced thus far for $R_3$, we see that linearity is by far the dominating concept, whether a vector is regarded in the algebraic sense as an ordered triple of numbers or in the geometric sense as a directed line segment extending from the origin. But in order to interpret algebraic vectors geometrically we drew on geometric experience or intuition to refer to such ideas as mutually perpendicular axes and the length of a directed segment. In Euclidean geometry, where length and angle are of central importance, these interpretations seem to be a natural means of visualizing algebraic concepts defined for the set of all ordered triples of real numbers. But if we consider an ordered triple $(a_1, a_2, a_3)$ of real numbers purely as an algebraic entity without any relation to familiar geometric systems, it is not apparent that we might also want to define the "length" of a triple or the "distance" and "angle" between two triples. Even if the desirability of having such definitions is clear, the precise form that the definitions should take is somewhat obscure. In the case of length there are some guidelines derived mainly from experience. We would agree, perhaps, that the length $\|(a_1, a_2, a_3)\|$ of the triple $(a_1, a_2, a_3)$ should be a nonnegative real number determined by the numbers $a_1, a_2, a_3$. We might even agree that $\|(0, 0, 0)\| = 0$ and $\|(a_1, a_2, a_3)\| > 0$ whenever at least one $a_i \neq 0$. But there are many reasonable choices available for the definition of length, such as

$$|a_1| + |a_2| + |a_3|,$$

$$(a_1^2 + a_2^2 + a_3^2)^{1/2},$$

$$(a_1^6 + a_2^6 + a_3^6)^{1/6},$$

$$\max(|a_1|, |a_2|, |a_3|).$$

Each of these definitions of length would provide an acceptable basis for making geometric interpretations of algebraic facts in $R_3$, but the corresponding geometry might not have the properties of Euclidean three-space. If we want to develop an algebraic structure for Euclidean geometry, we must adopt the Euclidean concepts of length, distance, and angle; we now see how this can be done for $R_3$.

We begin by insisting that length in $R_3$ be consistent with the customary notion of length in $R$; that is, the distance from the origin to the point on the real line that corresponds to the real number $b$ is $|b|$. In addition, we want the Pythagorean theorem to be valid in Euclidean three-space. These conditions require that the *length* $\|\alpha\|$ of a vector $\alpha = (a_1, a_2, a_3)$ be defined by

$$\|\alpha\| = \sqrt{a_1^2 + a_2^2 + a_3^2}.$$

Furthermore, each point in $R_3$ is the terminus of a vector, and the line segment from the terminus of $\alpha$ to the terminus of $\beta$ is a parallel translation of the vector $\beta - \alpha$; so $\|\beta - \alpha\|$ is simply the *distance* $d(\alpha, \beta)$ between the points $(a_1, a_2, a_3)$ and $(b_1, b_2, b_3)$.

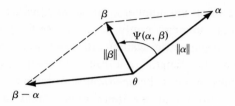

FIGURE 1.9

Thus

$$d(\alpha, \beta) = \|\beta - \alpha\| = \sqrt{(b_1 - a_1)^2 + (b_2 - a_2)^2 + (b_3 - a_3)^2},$$

$$\|\beta - \alpha\|^2 = (b_1^2 + b_2^2 + b_3^2) + (a_1^2 + a_2^2 + a_3^2) - 2(a_1 b_1 + a_2 b_2 + a_3 b_3),$$

$$\|\beta - \alpha\|^2 = \|\beta\|^2 + \|\alpha\|^2 - 2(a_1 b_1 + a_2 b_2 + a_3 b_3).$$

This equation strongly resembles the law of cosines, since $\alpha$ and $\beta$ are two sides of a triangle and $\|\beta - \alpha\|$ is the length of the third side. Since the law of cosines is valid in Euclidean space, we must have $a_1 b_1 + a_2 b_2 + a_3 b_3 = \|\alpha\| \, \|\beta\| \cos \Psi(\alpha, \beta)$, where $\Psi(\alpha, \beta)$ is the angle between vectors $\alpha$ and $\beta$. Hence

$$\cos \Psi(\alpha, \beta) = \frac{\sum\limits_{i=1}^{3} a_i b_i}{\left(\sum\limits_{i=1}^{3} a_i^2\right)^{1/2} \left(\sum\limits_{i=1}^{3} b_i^2\right)^{1/2}}.$$

This formula emphasizes the important role played in metric concepts by the scalar quantity $\sum\limits_{i=1}^{3} a_i b_i$, the sum of the products of corresponding components of two vectors. Recall from Exercises 6 of Section 1.1, that a similar expression formed the foundation for metric concepts in $R_2$. You are urged to review that exercise at this time, since the results stated there are in a form that is applicable also to $R_3$ and other Euclidean spaces.

In $R_3$ the *scalar product* $\alpha \cdot \beta$ of two vectors, where $\alpha = (a_1, a_2, a_3)$ and $\beta = (b_1, b_2, b_3)$, is defined by

$$\alpha \cdot \beta = a_1 b_1 + a_2 b_2 + a_3 b_3.$$

The scalar product is also called the *inner product* or *dot product*.

The scalar product $\alpha \cdot \beta$ is a function that assigns a real number to each pair of vectors in $R_3$. It is important to note that the scalar product is a linear function of each of its variables:

$$(a\alpha + b\beta) \cdot \gamma = a(\alpha \cdot \gamma) + b(\beta \cdot \gamma),$$
$$\alpha \cdot (b\beta + c\gamma) = b(\alpha \cdot \beta) + c(\alpha \cdot \gamma).$$

A function of two variables that is linear in each variable is called a *bilinear* function.

It is important also to observe that length, distance, and angle can be expressed in terms of the single concept of scalar product.

$$\|\alpha\| = (\alpha \cdot \alpha)^{1/2},$$
$$\|\beta - \alpha\|^2 = \|\alpha\|^2 + \|\beta\|^2 - 2\alpha \cdot \beta,$$
$$\alpha \cdot \beta = \|\alpha\| \, \|\beta\| \cos \Psi(\alpha, \beta).$$

We also know that $|\cos \Psi(\alpha, \beta)| \leq 1$, so we obtain

$$(\alpha \cdot \beta)^2 \leq (\alpha \cdot \alpha)(\beta \cdot \beta),$$

which is called the *Schwarz inequality*. Later in the book we shall derive these results in a more general setting.

The scalar product also provides a simple criterion for vectors to be *orthogonal* (perpendicular):

$\alpha$ and $\beta$ are orthogonal if and only if $\alpha \cdot \beta = 0$.

To construct a basis of mutually orthogonal vectors, we can use the *Gram-Schmidt orthogonalization process*. Begin with any nonzero vector $\alpha$ and any vector $\beta \notin [\alpha]$. As in Exercise 6 of Section 1.1, the orthogonal projection of $\beta$ upon $\alpha$ is the vector

$$\beta_1 = \frac{\alpha \cdot \beta}{\alpha \cdot \alpha} \alpha.$$

To demonstrate that this is so, we show that $\beta - \beta_1$ is orthogonal to $\alpha$, using the general properties of scalar product that are cited in Exercise 3 rather than laboring through the computation with coordinates. See Figure 1.10.

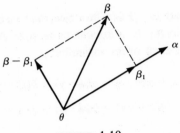

FIGURE 1.10

We have

$$(\beta - \beta_1) \cdot \alpha = \beta \cdot \alpha - \beta_1 \cdot \alpha = \beta \cdot \alpha - \left(\frac{\alpha \cdot \beta}{\alpha \cdot \alpha} \alpha\right) \cdot \alpha$$

$$= \beta \cdot \alpha - \frac{\alpha \cdot \beta}{\alpha \cdot \alpha} (\alpha \cdot \alpha) = \beta \cdot \alpha - \alpha \cdot \beta = 0,$$

as desired. Let $\beta_2 = \beta - \beta_1$. Then $\{\alpha, \beta_2\}$ is a linearly independent set of mutually orthogonal vectors. Extend this set to an orthogonal basis for $R_3$ by choosing any $\gamma \notin [\alpha, \beta_2]$, then subtracting from $\gamma$ its orthogonal projection on the plane $[\alpha, \beta_2]$. You may carry out the calculation to show that

$$\gamma_2 = \gamma - \frac{\alpha \cdot \gamma}{\alpha \cdot \alpha} \alpha - \frac{\beta_2 \cdot \gamma}{\beta_2 \cdot \beta_2} \beta_2$$

is orthogonal to both $\alpha$ and $\beta_2$. Then $\{\alpha, \beta_2, \gamma_2\}$ is an orthogonal basis for $R_3$.

If we want each basis vector to have unit length, as in the standard basis for $R_3$, we need only observe that if $\alpha$ is any nonzero vector, then the length of $\|\alpha\|^{-1}\alpha$ is 1. A basis of mutually orthogonal vectors of unit length is called a *normal orthogonal basis*, or an *orthonormal basis*.

To illustrate the Gram-Schmidt process, let us construct an orthogonal basis for $R_3$ beginning with the vector

$$\alpha = (2, -6, 3).$$

Now choose any vector $\beta \notin [\alpha]$. Actually it is easy to spot a vector that is already orthogonal to $\alpha$, such as

$$\beta_2 = (0, 1, 2),$$

avoiding the calculations inherent in the Gram-Schmidt process. To be sure that we have a third vector $\gamma \notin [\alpha, \beta_2]$ we rely on our knowledge that any vector in $[\alpha, \beta_2]$ is of the form

$$a\alpha + b\beta_2 = (2a, -6a + b, 3a + 2b).$$

If we pick specific values for $a$ and $b$, such as $a = 1$ and $b = 2$, to evaluate the first two components and then pick a third component *not* of the form $3a + 2b$, the resulting vector is not in $[\alpha, \beta_2]$. For example, choose $\gamma = (2, -4, 0)$. Then

$$\gamma_2 = \gamma - \frac{\alpha \cdot \gamma}{\alpha \cdot \alpha} \alpha - \frac{\beta_2 \cdot \gamma}{\beta_2 \cdot \beta_2} \beta_2$$

$$= (1, -2, 0) - \frac{14}{49}(2, -6, 3) - \frac{-2}{5}(0, 1, 2)$$

$$= \frac{1}{245}(105, 28, -14).$$

Then the set $\{\alpha, \beta_2, \gamma_2\}$ is mutually orthogonal, and a normal orthogonal basis for $R_3$ is given by

$$\left\{\frac{1}{7}\alpha, \frac{1}{\sqrt{5}}\beta_2, \frac{1}{\|\gamma_2\|}\gamma_2\right\}.$$

## EXERCISES 1.6

1. Apply the coordinate definition of length in $R_3$ to prove these general properties:

(i) $\|a\alpha\| = |a| \, \|\alpha\|$ for all $a \in R$, $\alpha \in R_3$;

(ii) $\|\alpha\| > 0$ if $\alpha \neq \theta$ and $\|\theta\| = 0$;

(iii) $\|\alpha + \beta\| \leq \|\alpha\| + \|\beta\|$.

Property (iii) is called the triangle inequality. Interpret it geometrically.

2. Prove that distance has these general properties:

(i) $d(\alpha, \beta) = d(\beta, \alpha)$;

(ii) $d(\alpha, \beta) > 0$ if $\alpha \neq \beta$, and $d(\alpha, \alpha) = 0$;

(iii) $d(\alpha, \beta) \leq d(\alpha, \gamma) + d(\gamma, \beta)$.

3. Prove that the scalar product in $R_3$ has these general properties:

(i) $(a\alpha + b\beta) \cdot \gamma = a(\alpha \cdot \gamma) + b(\beta \cdot \gamma)$;

(ii)  $\alpha \cdot \beta = \beta \cdot \alpha$;

(iii)  $\alpha \cdot \alpha > 0$ if $\alpha \neq \theta$.

4. Prove that

(i)  $\theta \cdot \eta = 0$ for every $\eta \in R_3$,

(ii) if $\xi \cdot \eta = 0$ for every $\eta \in R_3$, then $\xi = \theta$,

(iii) if $\xi_1 \cdot \eta = \xi_2 \cdot \eta$ for all $\eta \in R_3$, then $\xi_1 = \xi_2$.

5. Show that if $\{\alpha, \beta\}$ is linearly dependent, then the Schwarz inequality reduces to equality. Is the converse true? Prove your answer.

6. Prove that $(\alpha + \beta) \cdot (\alpha + \beta) + (\alpha - \beta) \cdot (\alpha - \beta) = 2\alpha \cdot \alpha + 2\beta \cdot \beta$, and interpret this result geometrically as a theorem about parallelograms.

7. Prove that $(\gamma - \alpha) \cdot (\gamma - \alpha) = (\beta - \alpha) \cdot (\beta - \alpha) + (\beta - \gamma) \cdot (\beta - \gamma)$ if and only if $(\beta - \alpha) \cdot (\beta - \gamma) = 0$, and interpret this result geometrically as a theorem about triangles.

8. In the notation of Euclidean three-space state and prove the theorem: the diagonals of a parallelogram are perpendicular if and only if the parallelogram is equilateral (a rhombus).

9. Each of the following triples of points in $R_3$ are the vertices of a triangle. Classify each triangle as equilateral, isosceles, right, or none of these. Use vector methods, and justify your conclusions.

(i) $A(1, 1, 1)$,   $B(3, -2, 2)$,   $C(4, 2, 1)$.
(ii) $A(1, 1, 1)$,   $B(0, 4, -1)$,   $C(-5, 3, 0)$.
(iii) $A(2, 3, -1)$,   $B(1, 5, 1)$,   $C(4, 2, 1)$.

10. Given a fixed vector $\gamma = (1, 2, 3)$ and an arbitrary vector $\xi = (x, y, z)$ in $R_3$.

(i) Write a scalar equation which expresses the condition that $\xi$ be orthogonal to $\gamma$.

(ii) Find two linearly independent vectors $\xi_1$ and $\xi_2$, both orthogonal to $\gamma$.

(iii) Show that every vector in the subspace $[\xi_1, \xi_2]$ is also orthogonal to $\gamma$.

(iv) Interpret these results geometrically.

11. Let $\gamma \neq \theta$ in $R_3$ and let $\gamma_1 = \|\gamma\|^{-1}\gamma$.

(i) Verify that $\gamma_1$ is a vector of unit length along $\gamma$.

(ii) Verify that for any vector $\alpha$, $(\alpha \cdot \gamma_1)\gamma_1$ is the orthogonal projection of $\alpha$ upon $\gamma$.

12. Given $\alpha = (-1, 2, 5)$ and $\beta = (3, -1, 1)$, find a unit vector $\gamma$ that is ortho-gonal to both $\alpha$ and $\beta$. Show also that $\gamma$ is orthogonal to every vector in the sub-space $[\alpha, \beta]$. Is $\{\alpha, \beta, \gamma\}$ a basis for $R_3$? An orthogonal basis?

13. Given $\alpha = (2, -2, 1)$ and $\beta = (1, 3, -1)$, answer the questions asked in Exercise 12.

14. Beginning with $\alpha = (2, 1, -2)$, use the Gram-Schmidt orthogonalization process to construct a normal orthogonal basis, the first vector of which is a multiple of $\alpha$.

15. Let $\{\alpha_1, \alpha_2, \alpha_3\}$ be a normal orthogonal basis for $R_3$, and let $\xi, \eta \in R_3$.

(i) Prove that $\xi \cdot \eta = \sum_{i=1}^{3} (\xi \cdot \alpha_i)(\eta \cdot \alpha_i)$.

(ii) Interpret geometrically the case where $\xi = \eta$.

## 1.7 Lines in $E_3$

The symbol $R_3$ has been used to denote the vector space of all real triples without reference to the scalar product or the associated metric concepts. Henceforth we shall use the symbol $E_3$ to denote the space of all real triples together with the usual scalar product; $E_3$ is called Euclid-ean three-space. The standard basis provides the usual form of repre-sentation for $E_3$, and in this section we shall develop more of its geometry. In this context it is conventional to use the standard basis denoted by the symbols $\bar{\imath}$, $\bar{\jmath}$, and $\bar{k}$ instead of $\epsilon_1$, $\epsilon_2$, $\epsilon_3$ respectively. Then the triple $(a_1, a_2, a_3)$ is written

$$a_1\bar{\imath} + a_2\bar{\jmath} + a_3\bar{k}.$$

When there is no need to exhibit the components of this vector, we shall continue to write $\alpha$, and we shall continue to think of $\alpha$ as a triple, a point in three-space, or a directed line segment from the origin — which-ever is most convenient.

From Exercise 11 of Section 1.6 we see that if $\gamma$ is a vector of unit length, then the perpendicular projection of a vector $\alpha$ on the line $[\gamma]$ is the vector $(\alpha \cdot \gamma)\gamma$. Specifically, if $\alpha = a_1\bar{\imath} + a_2\bar{\jmath} + a_3\bar{k}$, then

$$(\alpha \cdot \bar{\imath})\bar{\imath} = a_1\bar{\imath}, \qquad (\alpha \cdot \bar{\jmath})\bar{\jmath} = a_2\bar{\jmath}, \qquad (\alpha \cdot \bar{k})\bar{k} = a_3\bar{k}$$

are the projections of $\alpha$ on the coordinate axes. It follows that if $\alpha \neq \theta$,

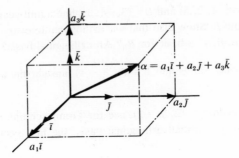

FIGURE 1.11

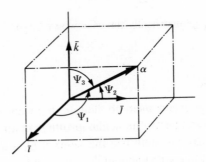

FIGURE 1.12

then the cosine of the angle $\Psi_1$ between the vectors $\alpha$ and $\bar{\imath}$ is given by

$$\cos \Psi_1 = \frac{\alpha \cdot \bar{\imath}}{\|\alpha\| \, \|\bar{\imath}\|} = \frac{a_1}{\|\alpha\|}.$$

Similarly, for $\bar{\jmath}$ and $\bar{k}$, we have $\cos \Psi_2 = \dfrac{a_2}{\|\alpha\|}$ and $\cos \Psi_3 = \dfrac{a_3}{\|\alpha\|}$. The numbers

$$\cos \Psi_1, \cos \Psi_2, \cos \Psi_3$$

are called the *direction cosines* of $\alpha$. Any nonzero multiple $k$ of the set of direction cosines of $\alpha$ is called a set of *direction numbers* of $\alpha$. By letting $k = \|\alpha\|$, we obtain $a_1, a_2, a_3$ as a particular set of direction numbers of $\alpha$. Furthermore, since any directed line segment in $E_3$ is a parallel translation of a uniquely determined vector emanating from the origin, we can define a set of direction numbers for that segment to be any set of direction numbers for that vector. In the same way, a set of direction numbers for

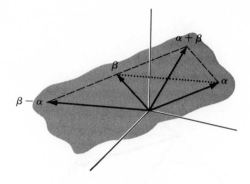

FIGURE 1.13

FIGURE 1.13

a line $L$ can be defined as a set of direction numbers for any directed segment of that line. Finally, we observe that if $a$, $b$, $c$ is a set of direction numbers for a line, then

$$\frac{a}{\sqrt{a^2 + b^2 + c^2}}, \quad \frac{b}{\sqrt{a^2 + b^2 + c^2}}, \quad \frac{c}{\sqrt{a^2 + b^2 + c^2}}$$

are the direction cosines of the vector which is a parallel translation to the origin of any segment of the line, suitably directed.

Recall that two vectors $\alpha$ and $\beta$ determine a parallelogram whose vertices are $\theta$, $\alpha$, $\beta$, and $\alpha + \beta$. This parallelogram is degenerate if and only if $\{\alpha,\ \beta\}$ is linearly dependent. The diagonal from the origin is $\alpha + \beta$. The other diagonal, represented by the directed segment from $\alpha$ to $\beta$, is a parallel translation of the vector $\beta - \alpha$, and the distance between $\alpha$ and $\beta$ is

$$\|\beta - \alpha\| = [(b_1 - a_1)^2 + (b_2 - a_2)^2 + (b_3 - a_3)^2]^{1/2}.$$

Now consider the representation of a line $L$ that has $c_1$, $c_2$, $c_3$ as a set of direction numbers and that passes through the point $(a_1, a_2, a_3)$. That is to say, the tip of the vector $\alpha = a_1\bar{\imath} + a_2\bar{\jmath} + a_3\bar{k}$ lies on $L$ and a suitable segment of the line is a parallel translation of the vector $\gamma = c_1\bar{\imath} + c_2\bar{\jmath} + c_3\bar{k}$. If $\xi = x\bar{\imath} + y\bar{\jmath} + z\bar{k}$ is a point on $L$, then for some scalar $t$, $\xi = \alpha + t\gamma$. As $t$ varies over $R$ every point of the line $L$ is generated (see Figure 1.14). Hence a parametric equation for $L$ can be written in vector form as

$$\xi = \alpha + t\gamma,$$

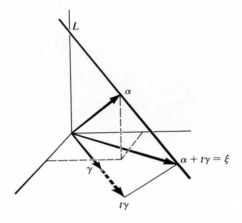

FIGURE 1.14

or, in terms of the components, as three parametric equations

$$x = a_1 + c_1 t,$$

$$y = a_2 + c_2 t,$$

$$z = a_3 + c_3 t.$$

If we solve for $t$ in each of these equations and equate the results, we obtain a Cartesian representation of $L$ in terms of the given direction numbers $c_i$ and the coordinates $a_i$ of the given point on $L$:

$$\frac{x - a_1}{c_1} = \frac{y - a_2}{c_2} = \frac{z - a_3}{c_3},$$

provided that $c_i \neq 0$ for each $i$.

If two points $\alpha = (a_1, a_2, a_3)$ and $\beta = (b_1, b_2, b_3)$ are given and we wish to find a representation of the line $L$ passing through $\alpha$ and $\beta$, we need note only that one set of direction numbers for $L$ is given by $b_1 - a_1$, $b_2 - a_2$, $b_3 - a_3$. From the previous calculations $L$ is represented parametrically by

$$\xi = \alpha + t(\beta - \alpha),$$

or in terms of coordinates by

$$\frac{x - a_1}{b_1 - a_1} = \frac{y - a_2}{b_2 - a_2} = \frac{z - a_3}{b_3 - a_3},$$

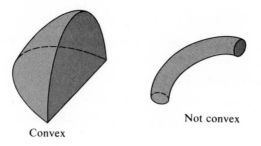

Convex

Not convex

FIGURE 1.15

provided that $b_i - a_i \neq 0$ for each $i$. Now $L$ is represented by two simultaneous linear equations in $x$, $y$, and $z$. Both here and in the calculations above, if one or two of the direction numbers of $L$ is zero, we still obtain for the representation of $L$ two simultaneous linear equations in $x$, $y$, and $z$, of which either one or two are of the form $x = a_1$, $y = a_2$, or $z = a_3$. Soon we shall see that a linear equation in $x$, $y$, and $z$ represents a plane in $E_3$, so two simultaneous linear equations represent $L$ as the set of points that are common to two intersecting planes.

Observe that a parametric equation for $L$ can be written

$$\xi = (1 - t)\alpha + t\beta;$$

every point of $L$ is therefore a linear combination of $\alpha$ and $\beta$ in which the sum of the two coefficients is 1:

$$\xi = a\alpha + b\beta \qquad \text{where } a + b = 1.$$

If $t = 1$, then $a = 0$ and $\xi = \beta$; if $t = 0$, then $a = 1$ and $\xi = \alpha$. For $0 \leq t \leq 1$ the points obtained are precisely those of the segment of $L$ from $\alpha$ to $\beta$. This observation is important in the study of convex sets. A subset $C$ of points of $E_3$ is said to be *convex* if and only if, whenever $\alpha$ and $\beta$ are in $C$, *every* point of the line segment between $\alpha$ and $\beta$ is also in $C$. It is not difficult to prove that a plane separates $E_3$ into two disjoint convex sets, that the intersection of any family of convex sets is convex, and that any real valued linear function defined on a closed convex set $C$ must assume its maximum and minimum values on $C$ at a boundary point of $C$. The last fact is important in the theory of linear programming.

## EXERCISES 1.7

1. A line $L$ passes through the point $A(3, 0, -1)$ and contains a parallel translation of the vector $2\bar{\imath} + 4\bar{\jmath} - 5\bar{k}$.

   (i)  Write a Cartesian equation for $L$.

   (ii) Determine the direction cosines of $L$.

2. Let $L$ be the line having direction numbers $-4, 3, -2$ and passing through the point $A(0, -2, 1)$.

   (i)   Write a vector equation for $L$.

   (ii)  Write component parametric equations for $L$.

   (iii) Write a Cartesian equation for $L$.

3. Given the points $A(-2, 1, 1)$ and $B(-3, 5, 4)$.

   (i)  Find direction cosines of the line $L$ through $A$ and $B$.

   (ii) Write a parametric equation for $L$ and also the Cartesian form of that equation.

   (iii) Determine the three points at which $L$ intersects the coordinate planes.

   (iv) Determine that point on $L$ that is closest to the origin.

4. Prove that $(\alpha + \beta) \cdot (\alpha + \beta) = \alpha \cdot \alpha + \beta \cdot \beta$ if and only if $\alpha \cdot \beta = 0$; interpret this result as a theorem in plane geometry.

5. Prove that $\|\alpha\| = \|\beta\|$ if and only if $(\alpha + \beta) \cdot (\alpha - \beta) = 0$; interpret this result as a theorem in plane geometry.

6. Prove that $\alpha$ and $\beta$ in $E_3$ are orthogonal if and only if $\|\alpha + t\beta\| \geq \|\alpha\|$ for every real number $t$.

7. Prove that

   (i)  if $A$ and $B$ are convex sets in $E_3$, then $A \cap B$ is also convex,

   (ii) the intersection of any family (not necessarily finite) of convex sets in $E_3$ is also convex.

## 1.8   Planes in $E_3$

Previously we discussed planes that are spanned by a linearly independent set $\{\alpha, \beta\}$ of two vectors. Any parallel translation of a plane through the origin is also called a plane. In short, a set $S$ of vectors of $E_3$ is called a *plane* if and only if for any $\sigma_0 \in S$ the set $S_0 = \{\sigma - \sigma_0 \mid \sigma \in S\}$ is a subspace that is spanned by two linearly independent vectors. It is

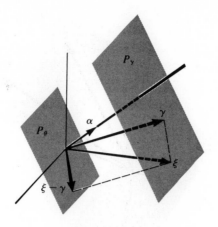

FIGURE 1.16

not difficult to verify that for any $\alpha \neq \theta$ the set of all vectors $\xi$ that are orthogonal to $\alpha$ is a plane $P_\theta$ through the origin having the vector equation

$$\alpha \cdot \xi = 0.$$

(A specific example was given in Exercise 10 of Section 1.6.) The parallel translation of $P_\theta$ by a fixed vector $\gamma$ is a plane $P_\gamma$ through $\gamma$ parallel to $P_\theta$, and every point $\xi$ of $P_\gamma$ will satisfy the equation

$$\alpha \cdot (\xi - \gamma) = 0,$$

$$\alpha \cdot \xi = \alpha \cdot \gamma.$$

In Cartesian form we have

$$a_1(x - c_1) + a_2(y - c_2) + a_3(z - c_3) = 0$$

as an equation of the plane $P_\gamma$ through the point $(c_1, c_2, c_3)$ and perpendicular to the direction $(a_1, a_2, a_3)$. This equation can also be written in the form

$$a_1 x + a_2 y + a_3 z = d,$$

where $d = \alpha \cdot \gamma = a_1 c_1 + a_2 c_2 + a_3 c_3$. But from the Gram-Schmidt orthogonalization process, we recall that the orthogonal projection of $\gamma$ on $\alpha$ is the vector

$$\left( \frac{\alpha \cdot \gamma}{\alpha \cdot \alpha} \right) \alpha = \frac{d}{\|\alpha\|} \frac{\alpha}{\|\alpha\|}.$$

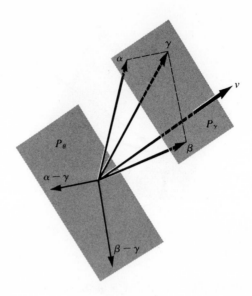

FIGURE 1.17

Hence $|d|(a_1^2 + a_2^2 + a_3^2)^{-1/2}$ is the perpendicular distance from the plane to the origin.

Conversely, the graph of the linear equation

$$ax + by + cz = d$$

is a plane $P$ perpendicular to a vector having $(a, b, c)$ as direction numbers and intersecting that vector at a distance $|d|(a^2 + b^2 + c^2)^{-1/2}$ from the origin. The vector $(a^2 + b^2 + c^2)^{-1/2} (a\bar{\imath} + b\bar{\jmath} + c\bar{k})$ is a vector of unit length perpendicular to the plane $P$; it is called the *unit normal* vector for $P$.

Another way of describing a plane is to specify three noncollinear points of $E_3$: $\alpha$, $\beta$, and $\gamma$. Then the set $\{\alpha - \gamma, \beta - \gamma\}$ is linearly independent, and $[\alpha - \gamma, \beta - \gamma]$ is a plane through the origin (see Figure 1.17). The translation of this plane by $\gamma$ produces a plane through $\alpha$, $\beta$, and $\gamma$, and a parametric equation of the latter plane is

$$\xi = s(\alpha - \gamma) + t(\beta - \gamma) + \gamma$$
$$= s\alpha + t\beta + (1 - s - t)\gamma \qquad \text{for all } s, t \in R.$$

The corresponding Cartesian equation can be obtained by eliminating the parameters $s$ and $t$ from the three linear equations in $x$, $y$, $z$, $s$, and $t$

that express this vector equation in terms of the components of $\alpha$, $\beta$, and $\gamma$.

Another method of writing a Cartesian equation for the plane through the points $\alpha$, $\beta$, and $\gamma$ is to find direction numbers $(n_1, n_2, n_3)$ of a vector $\nu$ normal to $P$. Then the desired equation is

$$n_1(x - c_1) + n_2(y - c_2) + n_3(z - c_3) = 0.$$

In the next section we shall find a convenient way to determine $\nu$ perpendicular to $(\alpha - \gamma)$ and $(\beta - \gamma)$.

Given a function $f$ of two variables, the graph of the equation

$$z = f(x, y)$$

is a surface in $E_3$ consisting of all points $P(x, y, f(x, y))$ for $(x, y)$ in the domain of $f$. Assuming that there exists a plane that is tangent to the surface at a point $P_0$, we can determine that tangent plane by reasoning as follows. Let $\xi = (x, y, z)$ be an arbitrary point on the plane, and let $\xi_0 = (x_0, y_0, z_0)$ be the point $P_0$, where $z_0 = f(x_0, y_0)$. Then the set of all points $\xi - \xi_0$ is a plane through the origin and is therefore the set of all linear combinations of any two independent vectors in the translated plane. We can guarantee independence by choosing one vector parallel to the surface in the $x$-direction ($y$ held constant) and the other vector parallel to the surface in the $y$-direction ($x$ held constant). Direction numbers for these vectors are

$$(1, 0, f_x(x_0, y_0)),$$

$$(0, 1, f_y(x_0, y_0)),$$

where $f_x$ and $f_y$ denote the partial derivatives of $f$ with respect to $x$ and $y$ respectively. Hence a vector equation of the tangent plane is

$$(x, y, z) - (x_0, y_0, z_0) = s(1, 0, f_x(x_0, y_0)) + t(0, 1, f_y(x_0, y_0)).$$

Then

$$x - x_0 = s,$$

$$y - y_0 = t,$$

so a scalar equation of the tangent plane is

$$z - z_0 = f_x(x_0, y_0)(x - x_0) + f_y(x_0, y_0)(y - y_0).$$

## EXERCISES 1.8

1. Given the vectors (or points)

$$\alpha = (-5, 5, 4),$$
$$\beta = (\ 4, 5, 1),$$
$$\gamma = (-4, 3, 0).$$

   (i)  Determine the angles that $\alpha$ makes with the coordinate axes.

   (ii)  Determine the largest angle of the triangle whose vertices are $\alpha$, $\beta$, and $\gamma$.

   (iii)  Write a parametric representation of the line through $\beta$ in the direction of $\alpha$. Represent the same line in Cartesian form.

   (iv)  Write an equation for the plane through $\beta$ and normal to the direction of $\gamma$.

   (v)  Write an equation of the plane through $\alpha$, $\beta$, and $\gamma$, and determine a unit normal for that plane.

   (vi)  Determine the distance from $\theta$ to the plane in (iv).

   (vii)  Determine the distance from $\alpha$ to the plane in (iv).

2. In $E_3$ let $P$ be the plane spanned by $\alpha = (1, -1, 2)$ and $\beta = (1, -1, -1)$. Given the vector $\gamma = (1, 5, 3)$, determine two vectors $\gamma_1$ and $\gamma_2$ such that $\gamma_1$ lies in the plane $P$, $\gamma_2$ is orthogonal to $P$, and $\gamma = \gamma_1 + \gamma_2$.

3. Write an equation for the tangent plane to the surface $z = f(x, y)$ at the point $P$, given that

   (i)  $f(x, y) = 4x^2 + y^2$, $P(1, -2, 8)$;

   (ii)  $f(x, y) = \sin (xy)$, $P(1, \pi, 0)$.

4. Given any plane $ax + by + cz = d$ in $E_3$, the set of points that satisfy $ax + by + cz \geq d$ is called a *closed half-space*, and the set of points that satisfy $ax + by + cz > d$ is called an *open half-space*. The same terminology is used when each inequality is reversed. Prove that each type of half-space is convex (see Exercise 7 of Section 1.7).

5. (i)  Let $f$ be a real linear function on $E_3$. $f(x, y, z) = px + qy + rz$. Prove that $f$ is a monotone function along every line of $E_3$; that is, along each line of $E_3$ either the values of $f$ remain constant or the values of $f$ steadily increase (or decrease, according to the direction).

   (ii)  Deduce that for each nonempty, closed, and bounded convex set $C$ and each linear function $f$, the maximum and minimum values of $f$ on $C$ occur on the boundary of $C$.

6. Sketch the region $C$ of $E_3$ that is described by the linear inequalities $2x + 3y + 4z \leq 12$, $x \geq 0$, $y \geq 0$, $z \geq 0$. Combine the conclusions of Exercises 4 and 5 to determine the maximum and minimum values of $f$ on $C$, where $f(x, y, z) = x - y + 2z$, and identify all points at which extreme values occur.

## 1.9   The Vector Product in $E_3$

As we have seen, the scalar or dot product in $E_3$ is a scalar valued function defined for each pair of vectors $\alpha = a_1\bar{i} + a_2\bar{j} + a_3\bar{k}$ and $\beta = b_1\bar{i} + b_2\bar{j} + b_3\bar{k}$ by

$$\alpha \cdot \beta = a_1b_1 + a_2b_2 + a_3b_3.$$

A second type of product $\alpha \times \beta$ can also be defined for each pair of vectors of $E_3$ such that $\alpha \times \beta$ turns out to be a *vector*. Hence this product is a vector valued function defined for each pair of vectors and is called the *vector* or *cross* product:

$$\alpha \times \beta = (a_2b_3 - a_3b_2)\bar{i} + (a_3b_1 - a_1b_3)\bar{j} + (a_1b_2 - a_2b_1)\bar{k}.$$

This definition stems from the work of W. R. Hamilton and was applied by Willard Gibbs to electromagnetic theory late in the nineteenth century. The geometric idea behind the definition is that $\alpha \times \beta$ is the vector that

(1)   is perpendicular to both $\alpha$ and $\beta$,

(2)   is of length $\|\alpha\| \, \|\beta\| \sin \Psi(\alpha, \beta)$,

(3)   extends from the origin in the direction of travel of a right-handed screw when it is rotated through the angle $\Psi(\alpha, \beta)$ *from $\alpha$ to $\beta$*, where $0 \leq \Psi(\alpha, \beta) \leq \pi$.

If $\{\alpha, \beta\}$ is linearly independent, Condition (1) determines a unique line $L$ through the origin, of which $\alpha \times \beta$ is a segment; Condition (2) determines the length of that segment; Condition (3) determines its direction along $L$. If $\{\alpha, \beta\}$ is linearly dependent, Condition (2) specifies $\alpha \times \beta = \theta$. To verify that the component definition of $\alpha \times \beta$ actually satisfies the three geometric conditions is a lengthy exercise in algebraic computation.

Similarly the following five algebraic properties of cross product can be verified for all $\alpha, \beta, \gamma \in E_3$ and all $c \in R$:

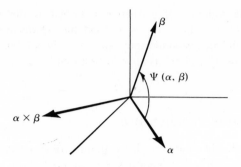

FIGURE 1.18

$$\alpha \times \beta = - \beta \times \alpha,$$

$$(\alpha + \beta) \times \gamma = (\alpha \times \gamma) + (\beta \times \gamma),$$

$$(c\alpha) \times \beta = c(\alpha \times \beta),$$

$$\alpha \cdot (\alpha \times \beta) = 0,$$

$$\|\alpha \times \beta\|^2 = (\alpha \cdot \alpha)(\beta \cdot \beta) - (\alpha \cdot \beta)^2.$$

The first property is one of anticommutativity or skew symmetry. The second and third properties state that $\alpha \times \beta$ is a linear function of $\alpha$, and it follows from skew symmetry that $\alpha \times \beta$ is also a linear function of $\beta$. Hence $\alpha \times \beta$ is bilinear. The fourth property is a restatement of Condition (1). From Condition (2) we see that the length of $\alpha \times \beta$ is numerically the same as the area of the parallelogram having $\alpha$ and $\beta$ as adjacent sides. The fifth property follows from combining this observation and the result of Exercise 6(v) of Section 1.1. Observe its relation to the Schwarz inequality.

The scalar product of $\alpha \times \beta$ with $\gamma = c_1 \bar{\imath} + c_2 \bar{\jmath} + c_3 k$ is easily computed:

$$(\alpha \times \beta) \cdot \gamma = (a_2 b_3 - a_3 b_2)c_1 + (a_3 b_1 - a_1 b_3)c_2 + (a_1 b_2 - a_2 b_1)c_3.$$

If you are familiar with determinants, you will recognize that the right hand side of this equation is the expansion of the three-by-three determinant

$$\det \begin{pmatrix} a_1 \ a_2 \ a_3 \\ b_1 \ b_2 \ b_3 \\ c_1 \ c_2 \ c_3 \end{pmatrix}.$$

If you have not studied determinants previously, you should learn this expansion now; we shall study n-by-n determinants in Chapter 4. For the present it is enough to observe that the value of the given three-by-three determinant is the sum of six terms, each of the form

$$\pm a_i b_j c_k.$$

The subscripts are the numbers 1, 2, and 3, taken in each of the six possible permutations. The + sign is attached to each term in which the subscripts appear *cyclicly* in the natural order: (1, 2, 3), or (2, 3, 1), or (3, 1, 2). The $-$ sign is used when the subscripts can be put into natural cyclic order by interchanging one pair of adjacent subscripts: (1, 3, 2), or (2, 1, 3), or (3, 2, 1).

Thus, determinants provide a convenient way of writing the vector product as a formal expression,

$$\alpha \times \beta = \det \begin{pmatrix} \bar{\imath} & \bar{\jmath} & \bar{k} \\ a_1 & a_2 & a_3 \\ b_1 & b_2 & b_3 \end{pmatrix}.$$

Likewise, the *scalar triple product* $(\alpha \times \beta) \cdot \gamma$ can be written

$$(\alpha \times \beta) \cdot \gamma = \det \begin{pmatrix} a_1 & a_2 & a_3 \\ b_1 & b_2 & b_3 \\ c_1 & c_2 & c_3 \end{pmatrix}.$$

It follows from properties of determinants that

$$(\alpha \times \beta) \cdot \gamma = \alpha \cdot (\beta \times \gamma).$$

Furthermore, from the properties of the scalar product, we see that

$$(\alpha \times \beta) \cdot \gamma = \|\alpha \times \beta\| \, \|\gamma\| \cos \Psi,$$

where $\Psi$ is the angle between $\alpha \times \beta$ and $\gamma$. But $\alpha \times \beta$ is perpendicular to the plane $[\alpha, \beta]$, so the absolute value of $\|\gamma\| \cos \Psi$ is the length of the component of $\gamma$ perpendicular to the plane $[\alpha, \beta]$. See Figure 1.19. Recalling that $\|\alpha \times \beta\|$ is the area of the parallelogram having $\alpha$ and $\beta$ as concurrent edges, we see that $|(\alpha \times \beta) \cdot \gamma|$ is the *volume of the parallelepiped* having $\alpha$, $\beta$, and $\gamma$ as concurrent edges.

From the foregoing we conclude that the absolute value of a three-by-three determinant is the volume of the parallelepiped having as

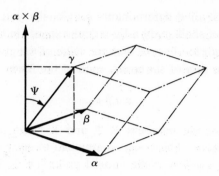

αⅹβ

FIGURE 1.19

concurrent edges the vectors whose components form the rows of that determinant.

Consider again the problem of writing an equation for the plane $P$ passing through three noncollinear points in space. If $\alpha$, $\beta$, and $\gamma$ are vectors from the origin to those points, then $(\alpha - \gamma) \times (\beta - \gamma)$ is a vector perpendicular to $P$. An equation for $P$ is

$$n_1(x - c_1) + n_2(y - c_2) + n_3(z - c_3) = 0,$$

where $(n_1, n_2, n_3)$ are components (or direction numbers) of $(\alpha - \gamma) \times (\beta - \gamma)$.

Unlike the other concepts that we have studied for $E_3$, the idea of the vector product does not generalize to vector spaces of every dimension. Indeed, it is a curious fact that a vector-valued function of pairs of vectors can be defined in Euclidean $n$-space to have the five properties listed earlier *only* for $n = 0, 1, 3$, and $7$; moreover, for $n = 0$ and $n = 1$ the geometry is wholly uninteresting. You can find a discussion of this result in a paper by Bertram Walsh, entitled "The Scarcity of Cross Products on Euclidean Spaces," *The American Mathematical Monthly*, **74**, 188–194 (1967). For interesting accounts of the historical development of vector algebra see E. T. Bell, *The Development of Mathematics*; New York: McGraw-Hill (1940); especially Chapters 8 and 9. See also the chapter by C. W. Curtis entitled "The Four and Eight Square Problem and Division Algebras," in *Studies in Modern Algebra*, A. A. Albert, editor; New York: Prentice-Hall (1963).

**EXERCISES** 1.9

1. Compute a table showing each of the nine vector products $\alpha \times \beta$ obtained by choosing $\alpha$ and $\beta$ to be $\bar{\imath}, \bar{\jmath}$, or $\bar{k}$ in all combinations.

2. Reason geometrically that $\alpha \times (\beta \times \gamma) \in [\beta, \gamma]$. Deduce from this that vector product is not an associative operation.

3. Given $\alpha = (-2, 1, 1)$, $\beta = (-3, 5, 4)$, and $\gamma = (3, 2, 1)$, compute $\alpha \times \beta$, $\beta \times \alpha$, $\alpha \times (\beta \times \gamma)$, and $(\alpha \times \beta) \times \gamma$.

4. Prove that $\alpha \cdot (\beta \times \gamma) = 0$ if and only if $\{\alpha, \beta, \gamma\}$ is linearly dependent.

5. Compute the volume of the parallelepiped having as concurrent edges the three vectors of Exercise 3. Interpret your answer.

6. Write an equation of the plane passing through the three points given in Exercise 3.

7. Prove that $(\alpha + b\beta) \times \beta = \alpha \times \beta$ for every real number $b$. Interpret this result as the area of various parallelograms to show geometrically why the equation is correct.

8. Use the component definition of $\alpha \times \beta$ to deduce the five algebraic properties listed in this section.

9. From Exercise 2, write $\alpha \times (\beta \times \gamma) = b\beta + c\gamma$. Compute the scalar product of each side with $\alpha$, to deduce that

$$\alpha \times (\beta \times \gamma) = k[(\alpha \cdot \gamma)\beta - (\alpha \cdot \beta)\gamma],$$

where $k$ is a real number independent of $\alpha$, $\beta$, and $\gamma$. (The fact that $k = 1$ can be shown by a rather long computation, which you need not carry out.)

10. Use the result

$$\alpha \times (\beta \times \gamma) = (\alpha \cdot \gamma)\beta - (\alpha \cdot \beta)\gamma$$

of Exercise 9 to derive the Jacobi identity:

$$\alpha \times (\beta \times \gamma) + \beta \times (\gamma \times \alpha) + \gamma \times (\alpha \times \beta) = \theta.$$

# 1.10   Linear Transformations in $R_3$

The calculations carried out in Section 1.5 illustrate the change of coordinates that occurs when one ordered basis $\{\alpha_1, \alpha_2, \alpha_3\}$ is changed to another ordered basis $\{\beta_1, \beta_2, \beta_3\}$, $\alpha_1$ being replaced by $\beta_1$, $\alpha_2$ by $\beta_2$, and $\alpha_3$ by $\beta_3$. We can interpret this process of replacement as a function $f$ that associates $\alpha_j$ with $\beta_j$ for $j = 1, 2, 3$:

$$\alpha_j = f(\beta_j) = \sum_{i=1}^{3} c_{ij}\beta_i.$$

Furthermore, the domain of $f$ can be extended from the set $\{\beta_1, \beta_2, \beta_3\}$ to all of $R_3$ by requiring that $f$ preserve linear combinations: that is, if

$$\xi = \sum_{i=1}^{3} b_i\beta_i,$$

then

$$f(\xi) = f\left(\sum_{i=1}^{3} b_i\beta_i\right) = \sum_{i=1}^{3} b_i f(\beta_i) = \sum_{i=1}^{3} b_i\alpha_i.$$

In this interpretation a vector $\xi$ with coordinates $[b_1, b_2, b_3]$ in the $\beta$-system is associated with the vector $f(\xi)$ having the same coordinates $[b_1, b_2, b_3]$ in the $\alpha$-system. In general, the two geometric vectors $\xi$ and $f(\xi)$ are distinct. Hence $f$ can be regarded as a transformation that carries each vector of $R_3$ into a vector of $R_3$. It is easy to verify that the function $f$ defined in this way satisfies the two properties

(a)   $f(\xi + \eta) = f(\xi) + f(\eta)$      for all $\xi, \eta \in R_3$,

(b)   $f(c\xi) = cf(\xi)$      for each $\xi \in R_3$ and each $c \in R$.

These two conditions simply restate the desired property that $f$ preserve all linear combinations of vectors of $R_3$, so $f$ is called a *linear transformation* of $R_3$. To distinguish linear transformations from other functions defined on $R_3$, we shall denote linear transformations with boldface capital letters such as **T**, **S**, and so on.

Before proceeding, we observe that the concept of a linear transformation does not require $\{\alpha_1, \alpha_2, \alpha_3\}$ to be a basis. For a given basis $\{\beta_1, \beta_2, \beta_3\}$ we can select *any* ordered triple of vectors $\{\alpha_1, \alpha_2, \alpha_3\}$ of $R_3$ and obtain a transformation **T** by defining

$$\mathbf{T}(\beta_j) = \alpha_j \qquad \text{for } j = 1, 2, 3.$$

If **T** is to be linear, the value of **T** at an arbitrary vector $\xi = \sum_{i=1}^{3} b_i\beta_i$ must be given by

$$\mathbf{T}(\xi) = \mathbf{T}\left(\sum_{i=1}^{3} b_i\beta_i\right) = \sum_{i=1}^{3} b_i\mathbf{T}(\beta_i) = \sum_{i=1}^{3} b_i\alpha_i.$$

A *linear transformation* **T** on $R_3$ is a function whose domain is $R_3$

and whose range is a subset of $R_3$ such that for all $\xi$, $\eta \in R_3$ and all $c, d \in R$,

$$\mathbf{T}(c\xi + d\eta) = c\mathbf{T}(\xi) + d\mathbf{T}(\eta).$$

Since linear transformations are functions, equality of linear transformations is defined as it is for functions: $\mathbf{T}_1 = \mathbf{T}_2$ if and only if, for each $\xi \in R_3$, $\mathbf{T}_1(\xi) = \mathbf{T}_2(\xi)$.

Linear transformations have many congenial properties, some of which you may prove as exercises. In particular we mention these:

(1)   A function $\mathbf{T}$ from $R_3$ to $R_3$ is linear if and only if, for all $\xi$, $\eta \in R_3$ and all $c \in R$, $\mathbf{T}(\xi + \eta) = \mathbf{T}(\xi) + \mathbf{T}(\eta)$ and $\mathbf{T}(c\xi) = c\mathbf{T}(\xi)$.

(2)   If $\mathbf{T}$ is linear, $\mathbf{T}(\theta) = \theta$.

(3)   If $\mathbf{T}$ is linear and if $S$ is a linearly dependent set, then $\mathbf{T}(S)$ is also linearly dependent, where $\mathbf{T}(S) = \{\mathbf{T}(\sigma) \mid \sigma \in S\}$.

(4)   Let $\mathbf{T}_1$ and $\mathbf{T}_2$ be linear transformations on $R_3$ and let $c \in R$; the sum $\mathbf{T}_1 + \mathbf{T}_2$, the scalar multiple $c\mathbf{T}_1$, and the composite $\mathbf{T}_2\mathbf{T}_1$ of these transformations are defined for all $\xi \in R_3$ by

$$(\mathbf{T}_1 + \mathbf{T}_2)(\xi) = \mathbf{T}_1(\xi) + \mathbf{T}_2(\xi),$$

$$(c\mathbf{T}_1)(\xi) = c(\mathbf{T}_1(\xi)),$$

$$(\mathbf{T}_2\mathbf{T}_1)(\xi) = \mathbf{T}_2(\mathbf{T}_1(\xi)).$$

Then each of the transformations $\mathbf{T}_1 + \mathbf{T}_2$, $c\mathbf{T}_1$, and $\mathbf{T}_2\mathbf{T}_1$ is also linear.

(5)   If $\mathbf{T}$ is linear, then $\mathbf{T}(R_3)$ is a subspace of $R_3$.

(6)   If $\mathbf{T}$ is linear, the set of all vectors $\alpha$ such that $\mathbf{T}(\alpha) = \theta$ is a subspace of $R_3$.

*Examples of Linear Transformations on $R_3$*

In the following examples the standard basis is used for geometric interpretations:

(a)   Let $\mathbf{T}$ be defined on $R_3$ by $\mathbf{T}(a, b, c) = (a, b, 0)$. $\mathbf{T}$ is the transformation that projects each point of three-space vertically onto the $x$-$y$ plane.
Either algebraically or geometrically it is easy to verify that $\mathbf{TT} = \mathbf{T}$, and hence that $\mathbf{T}^p = \mathbf{T}$ where, for every positive integer $p$, $\mathbf{T}^p = \mathbf{TT} \ldots \mathbf{T}$ ($p$ times). Such a transformation is called *idempotent*, meaning that all of its powers are the same. See Figure 1.20.

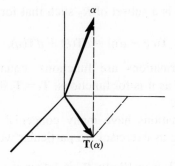

FIGURE 1.20

(b)   The *identity* transformation $\mathbf{I}$ is defined by $\mathbf{I}(a, b, c) = (a, b, c)$. Clearly $\mathbf{I}$ is linear and $\mathbf{TI} = \mathbf{T} = \mathbf{IT}$ for every transformation $\mathbf{T}$ of $R_3$.

(c)   The *zero* transformation $\mathbf{Z}$ is defined by $\mathbf{Z}(a, b, c) = (0, 0, 0)$. $\mathbf{Z}$ is linear and, for each $\mathbf{T}$, $\mathbf{T} + \mathbf{Z} = \mathbf{T} = \mathbf{Z} + \mathbf{T}$, and $\mathbf{TZ} = \mathbf{Z} = \mathbf{ZT}$.

(d)   The next example is more interesting. Let $\mathbf{T}$ be the linear transformation defined by

$$\mathbf{T}(\epsilon_1) = 2\epsilon_1 - 2\epsilon_2 + 3\epsilon_3,$$

$$\mathbf{T}(\epsilon_2) = \ \epsilon_1 + \ \epsilon_2 + \ \epsilon_3,$$

$$\mathbf{T}(\epsilon_3) = \ \epsilon_1 + 3\epsilon_2 - \ \epsilon_3.$$

The geometric effect of $\mathbf{T}$ is not at all obvious from these equations, but if we change bases in the proper way, the effect of $\mathbf{T}$ is easily described. Consider the effect of $\mathbf{T}$ on the vector $\alpha_1 = (5, 1, 4)$. $\mathbf{T}(\alpha_1) = 5\mathbf{T}(\epsilon_1) + \mathbf{T}(\epsilon_2) + 4\mathbf{T}(\epsilon_3) = 15\epsilon_1 + 3\epsilon_2 + 12\epsilon_3 = (15, 3, 12) = 3\alpha_1$. In the same manner, by choosing $\alpha_2 = (3, -5, 2)$ and $\alpha_3 = (0, 1, -1)$, we obtain $\mathbf{T}(\alpha_2) = \alpha_2$ and $\mathbf{T}(\alpha_3) = -2\alpha_3$. It is not hard to verify that $\{\alpha_1, \alpha_2, \alpha_3\}$ is a basis for $R_3$, so we can choose a new coordinate system, in which the coordinate axes are along $\alpha_1, \alpha_2, \alpha_3$. Then $\mathbf{T}$ can be described very simply by

$$\mathbf{T}(\alpha_1) = \ 3\alpha_1,$$

$$\mathbf{T}(\alpha_2) = \ \ \alpha_2,$$

$$\mathbf{T}(\alpha_3) = -2\alpha_3.$$

Geometrically $\mathbf{T}$ stretches each vector along the new first axis by a factor of 3; $\mathbf{T}$ leaves invariant each vector along the new second axis; vectors along the new third axis are stretched by a factor of 2 and also reflected through the origin. For an arbitrary vector $\xi = c_1\alpha_1 + c_2\alpha_2 + c_3\alpha_3$ we have $T(\xi) = 3c_1\alpha_1 + c_2\alpha_2 - 2c_3\alpha_3$.

Several questions should come to mind. In the general case how do we find vectors like $\alpha_1$, $\alpha_2$, $\alpha_3$ that are mapped in a simple manner by a given linear transformation? Do such vectors necessarily exist? Does a basis of such vectors exist? If not, how do we find a basis which represents $T$ as simply as possible? We shall return to these questions in Chapter 5.

## EXERCISES 1.10

1. Prove Properties (1–4) of linear transformations.

2. Determine whether each of the following functions from $R_3$ to $R_3$ is linear. If it is not linear, state why not.

$$f(a_1, a_2, a_3) = (0, 0, a_1 + a_2 + a_3),$$
$$g(a_1, a_2, a_3) = (a_2, a_1, 0),$$
$$h(a_1, a_2, a_3) = (a_1, a_2, 1),$$
$$j(a_1, a_2, a_3) = (a_2a_3, a_1a_3, a_1a_2),$$
$$k(a_1, a_2, a_3) = (0, 0, 0).$$

3. Prove Properties (5–6) of linear transformations.

4. Given $R_3$ with the standard basis, let $T$ be the linear transformation defined by

$$T(\epsilon_1) = (1, 2, 3),$$
$$T(\epsilon_2) = (2, 1, 3),$$
$$T(\epsilon_3) = (1, 1, 2).$$

Describe the space spanned by $\{T(\epsilon_1), T(\epsilon_2), T(\epsilon_3)\}$. Also determine all vectors $\alpha$ such that $T(\alpha) = \theta$.

5. Let $T$ be the linear transformation of $R_3$ defined by

$$T(1, 0, 0) = (1, 2, 3),$$
$$T(0, 1, 0) = (3, 1, 2),$$
$$T(0, 0, 1) = (2, 1, 3).$$

Compute $T(1, 1, 1)$, $T(3, -1, 4)$, and $T(4, 0, 5)$. Are these three vectors linearly independent?

6. Carry out the detailed computations that verify the assertions made in Example (d).

7. Property (3) of linear transformations asserts that a linear transformation preserves linear dependence. Find a specific example which shows that a linear transformation does not necessarily preserve linear independence.

8. Let **D** be the transformation of $R_3$ defined by $\mathbf{D}(a, b, c) = (0, 2a, b)$. Prove that **D** is linear. Show that $\mathbf{D}^2 \neq \mathbf{Z}$ but $\mathbf{D}^3 = \mathbf{Z}$, where **Z** is the zero linear transformation. (A transformation having this property is said to be *nilpotent* of index 3.) Interpret **D** as a linear transformation of the space of all polynomials of degree not exceeding 2 by associating the triple $(a, b, c)$ with the polynomial $ax^2 + bx + c$.

9. Consider the following linear transformations on the Cartesian plane $R_2$:

$$\mathbf{T}_1(a, b = (a \cos \Psi - b \sin \Psi, a \sin \Psi + b \cos \Psi)$$
$$\mathbf{T}_2(a, b) = (a, 0),$$
$$\mathbf{T}_3(a, b) = (0, b),$$
$$\mathbf{T}_4(a, b) = (b, a),$$
$$\mathbf{T}_5(a, b) = (a, b),$$
$$\mathbf{T}_6(a, b) = (0, 0).$$

(i) Verify that each transformation is linear, and describe the geometric effect of each.

(ii) Show that $\mathbf{T} + \mathbf{T}_6 = \mathbf{T}$ for every linear transformation **T** of $R_2$. $\mathbf{T}_6$ is therefore called the *zero* linear transformation **Z**.

(iii) Show that $\mathbf{T}\mathbf{T}_5 = \mathbf{T}_5\mathbf{T} = \mathbf{T}$ for every linear transformation **T** of $R_2$. $\mathbf{T}_5$ is therefore called the *identity* linear transformation **I**.

(iv) Show that $\mathbf{T}_2\mathbf{T}_3 = \mathbf{Z}$ although $\mathbf{T}_2 \neq \mathbf{Z}$ and $\mathbf{T}_3 \neq \mathbf{Z}$.

(v) Show that $\mathbf{T}_2\mathbf{T}_4 \neq \mathbf{T}_4\mathbf{T}_2$.

(vi) Show that $\mathbf{T}_2\mathbf{T}_2 = \mathbf{T}_2$ although $\mathbf{T}_2 \neq \mathbf{Z}$ and $\mathbf{T}_2 \neq \mathbf{I}$.

(vii) Show that if $\{\alpha, \beta\}$ is a basis, then $\{\mathbf{T}_1(\alpha), \mathbf{T}_1(\beta)\}$ is also a basis.

10. Let $\mathscr{T}$ denote the set of *all* linear transformations from $R_3$ to $R_3$. Sum and scalar multiple of linear transformations are defined as in Property (4). Show that with these operations $\mathscr{T}$ satisfies the eight axioms of a vector space, as listed in Section 1.2.

## 1.11  Matrix Algebra for $R_3$

We have already seen that matrices can be used to represent systems of linear equations and that the work of solving a system can be interpreted in terms of specified operations on the rows of a matrix. Now we investigate the use of matrices to represent linear transformations on $R_3$ and to systematize certain computations.

Let **T** be any linear transformation on $R_3$, and let $\{\alpha_1, \alpha_2, \alpha_3\}$ be an arbitrary but fixed basis of $R_3$. Each vector $\xi \in R_3$ can be represented uniquely as a linear combination of basis vectors,

$$\xi = x_1\alpha_1 + x_2\alpha_2 + x_3\alpha_3 = \sum_{j=1}^{3} x_j\alpha_j.$$

Since **T** is linear,

$$\mathbf{T}(\xi) = \mathbf{T}\left(\sum_{j=1}^{3} x_j\alpha_j\right) = \sum_{j=1}^{3} x_j\mathbf{T}(\alpha_j).$$

Hence the effect of **T** on each vector of $R_3$ is completely determined by the effect of **T** on the basis vectors. Suppose that

$$\mathbf{T}(\alpha_1) = a_{11}\alpha_1 + a_{21}\alpha_2 + a_{31}\alpha_3 = \sum_{i=1}^{3} a_{i1}\alpha_i,$$

$$\mathbf{T}(\alpha_2) = a_{12}\alpha_1 + a_{22}\alpha_2 + a_{32}\alpha_3 = \sum_{i=1}^{3} a_{i2}\alpha_i,$$

$$\mathbf{T}(\alpha_3) = a_{13}\alpha_1 + a_{23}\alpha_2 + a_{33}\alpha_3 = \sum_{i=1}^{3} a_{i3}\alpha_i.$$

Then

$$\mathbf{T}(\xi) = (x_1a_{11} + x_2a_{12} + x_3a_{13})\alpha_1$$
$$+ (x_1a_{21} + x_2a_{22} + x_3a_{23})\alpha_2$$
$$+ (x_1a_{31} + x_2a_{32} + x_3a_{33})\alpha_3$$
$$= \sum_{i=1}^{3} \left(\sum_{j=1}^{3} a_{ij}x_j\right)\alpha_i.$$

The conventions of matrix algebra are chosen in such a way that the study of linear transformations can be carried out algebraically with matrices.

We recall our earlier convention of representing a vector, relative to a chosen basis, by the *column* of its components or coordinates relative to that basis. Hence relative to the $\alpha$-basis the vector $\xi$ is represented in matrix form by a *column vector X*:

$$\xi \to X = \begin{pmatrix} x_1 \\ x_2 \\ x_3 \end{pmatrix}.$$

The linear transformation **T** is represented by the three-by-three array of scalars obtained by placing the components of $\mathbf{T}(\alpha_1)$ in the first column, the components of $\mathbf{T}(\alpha_2)$ in the second column, and the components of $\mathbf{T}(\alpha_3)$ in the third column to obtain a square *matrix A*:

$$\mathbf{T} \to A = \begin{pmatrix} a_{11} & a_{12} & a_{13} \\ a_{21} & a_{22} & a_{23} \\ a_{31} & a_{32} & a_{33} \end{pmatrix}.$$

Observe that the entry in row $i$ and column $j$ of $A$ is the real number $a_{ij}$. We shall frequently write

$$A = (a_{ij}) \qquad i, j = 1, 2, 3,$$

instead of writing the matrix in extended form.

Since a column vector can be regarded as the coordinates of a point in space, two column vectors are equal if and only if corresponding entries are equal. We use the analogous definition for equality of matrices:

If $A = (a_{ij})$ and $B = (b_{ij})$ are three-by-three matrices, then $A = B$ if and only if $a_{ij} = b_{ij}$, for $i = 1, 2, 3$ and $j = 1, 2, 3$.

From our earlier calculations, the vector $\mathbf{T}(\xi)$ is represented by a column vector $Y$:

$$\mathbf{T}(\xi) \to Y = \begin{pmatrix} y_1 \\ y_2 \\ y_3 \end{pmatrix}, \qquad \text{where } y_i = a_{i1}x_1 + a_{i2}x_2 + a_{i3}x_3.$$

In order to assure that matrix algebra reflects this representation, we define the *product* of a three-by-three matrix $A$ and a 3-component column vector $X$ by the rule

$$\begin{pmatrix} a_{11} & a_{12} & a_{13} \\ a_{21} & a_{22} & a_{23} \\ a_{31} & a_{32} & a_{33} \end{pmatrix} \begin{pmatrix} x_1 \\ x_2 \\ x_3 \end{pmatrix} = \begin{pmatrix} y_1 \\ y_2 \\ y_3 \end{pmatrix},$$

where

$$y_1 = a_{11}x_1 + a_{12}x_2 + a_{13}x_3,$$
$$y_2 = a_{21}x_1 + a_{22}x_2 + a_{23}x_3,$$
$$y_3 = a_{31}x_1 + a_{32}x_2 + a_{33}x_3.$$

Note that $y_i$ is the scalar product of the $i$th *row vector* of $A$ and the *column vector X*.

If **T** is represented by the three-by-three matrix $A$, and if $\xi$ is represented by the three-by-one matrix $X$, then $\mathbf{T}(\xi)$ is represented by the three-by-one matrix $AX$.

*Example*

$$\begin{pmatrix} 3 & 1 & -2 \\ 4 & 0 & 1 \\ -1 & 2 & -2 \end{pmatrix} \begin{pmatrix} -2 \\ 4 \\ 1 \end{pmatrix} = \begin{pmatrix} a_1 \\ a_2 \\ a_3 \end{pmatrix},$$

where

$$a_1 = (3)(-2) + (1)(4) + (-2)(1) = -4,$$

$$a_2 = (4)(-2) + (0)(4) + (1)(1) = -7,$$

$$a_3 = (-1)(-2) + (2)(4) + (-2)(1) = 8.$$

The rule for the *scalar multiple* of a matrix also is derived from consideration of a linear transformation. If **T** is a linear transformation represented by $A$, and if $k$ is a scalar, then since **T** is a function, so is $k\mathbf{T}$, and

$$(k\mathbf{T})(\xi) = k\mathbf{T}(\xi) \qquad \text{for every } \xi \text{ in } R_3.$$

Letting $\xi = \alpha_j$, we see that column $j$ of the matrix that represents $k\mathbf{T}$ is simply the product of the scalar $k$ and column $j$ of $A$, for each $j = 1,2,3$.

If $A = (a_{ij})$ is a three-by-three matrix and $k$ is a scalar, then $kA = (b_{ij})$, where $b_{ij} = ka_{ij}$ for $i, j = 1, 2, 3$.

*Example*

$$2 \begin{pmatrix} 3 & 1 & -2 \\ 4 & 0 & 1 \\ -1 & 2 & -2 \end{pmatrix} = \begin{pmatrix} 6 & 2 & -4 \\ 8 & 0 & 2 \\ -2 & 4 & -4 \end{pmatrix}.$$

Now let **T** and **S** be two linear transformations, represented relative to the chosen basis by the two three-by-three matrices $A$ and $B$. The *sum* of **T** and **S** is defined as is usual for functions,

$$(\mathbf{T} + \mathbf{S})(\xi) = \mathbf{T}(\xi) + \mathbf{S}(\xi) \qquad \text{for each } \xi \in R_3.$$

Hence for each $j$

$$(\mathbf{T} + \mathbf{S})(\alpha_j) = \mathbf{T}(\alpha_j) + \mathbf{S}(\alpha_j).$$

Thus the matrix that represents $\mathbf{T} + \mathbf{S}$ has as its $j$th column vector the *sum* of the $j$th column vectors of $A$ and $B$. Since the sum of two vectors is performed component-by-component, matrix addition must also be defined component-by-component.

If $A = (a_{ij})$ and $B = (b_{ij})$ are three-by-three matrices, then

$$A + B = (c_{ij}) \qquad \text{where } c_{ij} = a_{ij} + b_{ij} \text{ for } i, j = 1, 2, 3.$$

**Example**

$$\begin{pmatrix} 3 & 1 & -2 \\ 4 & 0 & 1 \\ -1 & 2 & -2 \end{pmatrix} + \begin{pmatrix} -1 & 2 & 2 \\ 0 & 1 & -1 \\ 3 & 1 & -2 \end{pmatrix} = \begin{pmatrix} 2 & 3 & 0 \\ 4 & 1 & 0 \\ 2 & 3 & -4 \end{pmatrix}.$$

Since both **T** and **S** are functions from $R_3$ to $R_3$, the composite function **TS** is defined,

$$(\mathbf{TS})(\xi) = \mathbf{T}(\mathbf{S}(\xi)) \qquad \text{for every } \xi \in R_3.$$

To derive a matrix representation for **TS** in terms of the matrices $A$ and $B$ that represent **T** and **S** respectively, we compute as follows:

$$(\mathbf{TS})(\alpha_j) = \mathbf{T}(\mathbf{S}(\alpha_j)) = \mathbf{T}\left(\sum_{k=1}^{3} b_{kj}\alpha_k\right)$$

$$= \sum_{k=1}^{3} b_{kj}\mathbf{T}(\alpha_k) = \sum_{k=1}^{3} b_{kj}\left(\sum_{i=1}^{3} a_{ik}\alpha_i\right)$$

$$= \sum_{i=1}^{3} \left(\sum_{k=1}^{3} a_{ik}b_{kj}\right)\alpha_i.$$

Hence column $j$ of the matrix $C$ that represents **TS** is the column vector whose $i$th component is $\sum_{k=1}^{3} a_{ik}b_{kj}$. Another way of describing $C$ is to observe that

$$c_{ij} = \sum_{k=1}^{3} a_{ik}b_{kj},$$

which is the scalar product of row $i$ of $A$ and column $j$ of $B$.

If $A = (a_{ij})$ and $B = (b_{ij})$ are three-by-three matrices, then $AB = (c_{ij})$, where $c_{ij} = \sum_{k=1}^{3} a_{ik}b_{kj}$ for $i, j = 1, 2, 3$.

**Example**

$$\begin{pmatrix} 3 & 1 & -2 \\ 4 & 0 & 1 \\ -1 & 2 & -2 \end{pmatrix} \begin{pmatrix} -1 & 2 & 2 \\ 0 & 1 & -1 \\ 3 & 1 & -2 \end{pmatrix} = \begin{pmatrix} -9 & 5 & 9 \\ -1 & 9 & 6 \\ -5 & -2 & 0 \end{pmatrix}.$$

The entry in row 1 and column 2 of the product is obtained by computing the scalar product of row 1 of the first matrix and column 2 of the second:

$$(3)(2) + (1)(1) + (-2)(1) = 5.$$

You should verify the other entries.

In Chapter 3 we shall consider matrix algebra more generally and more thoroughly, proving a number of properties that we merely state for three-by-three matrices at this time. Further properties are illustrated by the exercises.

(a)   Matrix addition is associative and commutative:

$$A + (B + C) = (A + B) + C,$$

$$A + B = B + A.$$

(b)   Scalar multiples of matrices satisfy

$$(a + b)A = aA + bA,$$

$$a(A + B) = aA + aB.$$

(c)   Matrix multiplication is associative and bilinear:

$$A(BC) = (AB)C$$

$$(aA + bB)C = aAC + bBC$$

$$A(bB + cC) = bAB + cAC.$$

By computing with the previous numerical example, you can readily verify that *matrix multiplication is not commutative*. It is interesting to note that Properties (a) and (b) above are precisely the same as Properties (1), (2), (5), and (6) of the eight basic properties of a vector space, listed in Section 1.2. Does the set of all three-by-three real matrices form a real vector space? As an exercise you may show that the remaining four properties are indeed satisfied, so the answer is *yes*.

## EXERCISES 1.11

Exercises 1–10 below refer to the following vectors and matrices.

$$X = \begin{pmatrix} x_1 \\ x_2 \\ x_3 \end{pmatrix}, \qquad Y = \begin{pmatrix} 0 \\ 1 \\ 2 \end{pmatrix}, \qquad E = \begin{pmatrix} 1 \\ 1 \\ 1 \end{pmatrix},$$

$$A = \begin{pmatrix} a_{11} & a_{12} & a_{13} \\ a_{21} & a_{22} & a_{23} \\ a_{31} & a_{32} & a_{33} \end{pmatrix}, L = \begin{pmatrix} 0 & 0 & 0 \\ 1 & 0 & 0 \\ 1 & 1 & 0 \end{pmatrix}, \quad D = \begin{pmatrix} 2 & 0 & 0 \\ 0 & 3 & 0 \\ 0 & 0 & 4 \end{pmatrix},$$

$$Z = \begin{pmatrix} 0 & 0 & 0 \\ 0 & 0 & 0 \\ 0 & 0 & 0 \end{pmatrix}, \quad I = \begin{pmatrix} 1 & 0 & 0 \\ 0 & 1 & 0 \\ 0 & 0 & 1 \end{pmatrix}, U_{21} = \begin{pmatrix} 0 & 0 & 0 \\ 1 & 0 & 0 \\ 0 & 0 & 0 \end{pmatrix}.$$

1. Compute $3D - 2L, LD, DL$.

2. Compute $LY$ and $AE$. Observe that the vector $AE$ describes the *row sums* of the matrix $A$.

3. Clearly $A + Z = A = Z + A$, so $Z$ is called the three-by-three *zero matrix*. Compute $AZ$ and $ZA$.

4. Verify that $AI = A = IA$. Hence $I$ is called the three-by-three *identity matrix*.

5. Compute $D^2$ and $D^3$. Matrices such as $D$ and $I$, in which each entry not on the main diagonal is zero, are called *diagonal matrices*.

6. Compute $DA$ and $AD$. Computation with diagonal matrices is considerably simpler than with general matrices.

7. Compute $L^2, L^3$. Since the nonzero elements of $L$ appear only below the main diagonal, $L$ is called a *lower triangular* matrix. $L$ is also said to be *nilpotent of index 3* since $L^3 = Z$ but $L^2 \neq Z$.

8. Compute $U_{21}A$ and $AU_{21}$, and describe the effect of $U_{21}$ as a left multiplier of $A$ and as a right multiplier of $A$.

9. Verify that $U_{21}L = Z$ but $LU_{21} \neq Z$.

10. Write out fully the system of linear equations represented by the matrix equation $LX = Y$. Solve the system.

11. Let $\mathcal{M}_{3 \times 3}$ denote the set of all three-by-three matrices of real numbers. Verify that $\mathcal{M}_{3 \times 3}$ satisfies the eight basic properties of a vector space.

12. Consider the nine three-by-three unit matrices $U_{rs}$ in which $u_{rs} = 1$ and all other entries are 0. (One of these is $U_{21}$.) Prove that these nine matrices form

a basis of the vector space $\mathscr{M}_{3\times3}$; that is, prove that any three-by-three matrix is a linear combination of those nine matrices, and that the set of those nine matrices is linearly independent.

## 1.12 Quadratic Forms in $E_3$

Having seen that matrices are useful in studying linear transformations and systems of linear equations, we should not be surprised to find applications of matrix theory in a variety of mathematical problems. A few examples are treated briefly in Section 1.13, and in this section we consider quadratic forms in three variables, a topic that we shall treat more generally in Chapter 5.

### Examples of Quadratic Forms

(a) The expression for the fundamental metric (element of arc length) in the Euclidean plane is

$$ds^2 = dx^2 + dy^2 \text{ in rectangular coordinates,}$$

$$ds^2 = dr^2 + r^2d\theta^2 \text{ in polar coordinates.}$$

(b) The fundamental metric in Euclidean three-space is

$$ds^2 = dx^2 + dy^2 + dz^2 \text{ in rectangular coordinates,}$$

$$ds^2 = dr^2 + r^2d\theta^2 + dz^2 \text{ in cylindrical coordinates,}$$

$$ds^2 = d\rho^2 + \rho^2\sin^2\Psi d\theta^2 + \rho^2 d\Psi^2 \text{ in spherical coordinates.}$$

(c) In classical mechanics the kinetic energy of a particle of mass $m$ and having $n$ degrees of freedom is given by

$$KE = \frac{m}{2}\sum_{i=1}^{n}\left(\frac{dx_i}{dt}\right)^2,$$

where the $x_i$ are the position coordinates of the particle. In more general systems the kinetic energy is represented by more complicated quadratic forms.

As background for our study of quadratic forms in $E_3$, let us recall what we know about the corresponding problem in $E_2$. The most general quadratic equation in two variables can be written

$$ax^2 + bxy + cy^2 + dx + ey + f = 0,$$

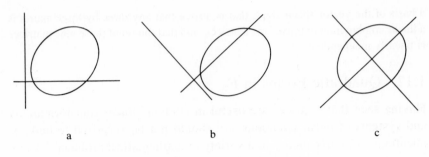

a                                b                                c

FIGURE 1.21

and its graph in $E_2$ is an ellipse, parabola, or hyperbola according to whether $b^2 - 4ac$ is negative, zero, or positive. See Figure 1.21,a. A suitable rotation of axes aligns the coordinate axes with the axes of the conic, as in Figure 1.21,b. Then a translation of axes carries the origin to the center of an ellipse or hyperbola or to the vertex of a parabola, as in Figure 1.21,c. Relative to this coordinate system the conic has an equation in simple standard form:

$$LX^2 + MY^2 = K \qquad \text{for an ellipse or hyperbola,}$$

$$Y = NX^2 \qquad \text{for a parabola.}$$

Note in particular that translations and rotations are distance preserving transformations (rigid motions) of the plane. A rotation is linear, but a translation is nonlinear since a linear transformation keeps the origin fixed.

For quadratic equations in $E_3$ we shall assume that the origin is properly positioned so that the equation contains no first degree terms:

$$ax^2 + 2bxy + cy^2 + 2dxz + 2eyz + fz^2 = k.$$

The graph of this equation is a central quadric surface in $E_3$, and in order to study its nature we seek a distance preserving change of coordinates that will bring the equation to the simplified form

$$LX^2 + MY^2 + NZ^2 = K.$$

The expression

$$q(x_1,x_2,x_3) = a_{11}x_1{}^2 + 2a_{12}x_1x_2 + a_{22}x_2{}^2$$
$$+ 2a_{13}x_1x_3 + 2a_{23}x_2x_3 + a_{33}x_3{}^2$$

is called a *quadratic form* in the three variables $x_1$, $x_2$, and $x_3$. The form can be represented by the matrix

$$A = \begin{pmatrix} a_{11} & a_{12} & a_{13} \\ a_{12} & a_{22} & a_{23} \\ a_{13} & a_{23} & a_{33} \end{pmatrix}$$

in which the $(i, j)$ entry specifies the coefficient of $x_i x_j$. Note that the term $2a_{12}x_1x_2$ is split into $a_{12}x_1x_2 + a_{12}x_2x_1$, causing $A$ to be symmetric across its main diagonal, from the upper left-hand corner to the lower right-hand corner.

A three-by-three matrix $B = (b_{ij})$ is said to be *symmetric* if and only if $b_{ij} = b_{ji}$ for all $i, j$.

Another way to describe symmetry of a matrix is to introduce the notion of the transpose of the matrix.

If $B = (b_{ij})$ is a three-by-three matrix, then the *transpose* $B^t$ of $B$ is the three-by-three matrix

$$B^t = (c_{ij}), \qquad \text{where } c_{ij} = b_{ji},$$

In effect, $B^t$ is obtained by using each column of $B$ as the corresponding row of $B^t$. Then $B$ is symmetric if and only if $B = B^t$. If we extend the idea of transpose to a column vector $X$, considered as a three-by-one matrix, then $X^t$ is a one-by-three matrix, or a row vector. If

$$X = \begin{pmatrix} x_1 \\ x_2 \\ x_3 \end{pmatrix}, \qquad \text{then } X^t = (x_1\ x_2\ x_3).$$

The product $X^t B$ of a three component row vector and a three-by-three matrix $B$ can be defined as for two matrices, where the entry in row $i$ and column $j$ of the product is simply the scalar product of row $i$ of the first matrix with column $j$ of the second:

$$X^t B = (x_1\ x_2\ x_3) \begin{pmatrix} b_{11} & b_{12} & b_{13} \\ b_{21} & b_{22} & b_{23} \\ b_{31} & b_{32} & b_{33} \end{pmatrix} = (c_1\ c_2\ c_3),$$

where

$$c_j = x_1 b_{1j} + x_2 b_{2j} + x_3 b_{3j}.$$

As an exercise you may verify that the transpose of a product of three-by-three matrices is the product of the transposes *in the reverse order.*

$$(BC)^t = C^t B^t.$$

Similarly, for a three-by-three matrix $B$ and a three-by-one matrix $X$,

$$(BX)^t = X^t B^t.$$

Now we return to the matrix representation of the quadratic form $q(x_1, x_2, x_3)$, defined previously. Clearly $q$ is a real valued function, defined at each point of $E_3$ and represented by the symmetric matrix $A$. If $\xi$ is a point whose coordinates are given by $X$ then $q(\xi)$ is given by the matrix product

$$q(\xi) = X^t A X,$$

as may be verified by carrying out this matrix computation.

Our problem is to find a new coordinate system for $R_3$, relative to which $q$ is represented by a diagonal matrix $D$, for if $\xi$ were represented in that coordinate system by the column vector $Y$ we would have

$$q(\xi) = Y^t D Y$$

$$= (y_1 \; y_2 \; y_3) \begin{pmatrix} d_1 & 0 & 0 \\ 0 & d_2 & 0 \\ 0 & 0 & d_3 \end{pmatrix} \begin{pmatrix} y_1 \\ y_2 \\ y_3 \end{pmatrix}$$

$$= d_1 y_1^2 + d_2 y_2^2 + d_3 y_3^2,$$

as desired. Furthermore, we require the linear transformation that performs the change of coordinates to be distance preserving (a rigid motion). The *Principal Axes Theorem*, which is important in various parts of physics and engineering, asserts the existence of a distance preserving change of coordinates such that a given quadratic form can be converted to a sum of squares. However, at this early stage of our study we are ill prepared to carry out a proof of this theorem, and we defer further consideration of it until we have developed a wider knowledge of linear algebra.

It will be instructive, however, to consider briefly the geometry of a linear transformation $T$ that can be represented relative to some basis $\{\alpha_1, \alpha_2, \alpha_3\}$ by a diagonal matrix $D$. Since the columns of $D$ specify the coordinates of $T(\alpha_1)$, $T(\alpha_2)$, and $T(\alpha_3)$ relative to the $\alpha$-basis we have

$$T(\alpha_1) = d_1\alpha_1 + 0\alpha_2 + 0\alpha_3,$$
$$T(\alpha_2) = 0\alpha_1 + d_2\alpha_2 + 0\alpha_3,$$
$$T(\alpha_3) = 0\alpha_1 + 0\alpha_2 + d_3\alpha_3.$$

Hence each basis vector is mapped by **T** into a scalar multiple of itself.

A nonzero vector $\xi$ is called a *characteristic vector* of **T** if and only if there exists a scalar $c$ such that

$$\mathbf{T}(\xi) = c\xi.$$

This scalar $c$ is called a *characteristic value* of **T** associated with the characteristic vector $\xi$.

We have already seen an illustration of characteristic vectors in Example (a) of Section 1.10, and the concept is a very useful tool in linear algebra and many of its applications.

## EXERCISES 1.12

1. Write a symmetric matrix that represents the given quadratic form:
   (i) $3x^2 - 2xy + y^2 - xz - z^2$,
   (ii) $xy + 2xz + yz$.

2. Verify the assertion in the text that $q(\xi) = X^t A X$.

3. Prove the following properties of the transpose for three-by-three matrices:
   (i) $(A^t)^t = A$,
   (ii) $(A + B)^t = A^t + B^t$,
   (iii) $(cA)^t = cA^t$,
   (iv) $(AB)^t = B^t A^t$.

4. Prove that for every three-by-three matrix $A$, the matrices $A + A^t$ and $AA^t$ are both symmetric. (Use Exercise 3.)

5. Let $A$ and $B$ be symmetric three-by-three matrices.
   (i) Prove that $A + B$ is symmetric.
   (ii) Find a necessary and sufficient condition that $AB$ be symmetric.

6. A three-by-three matrix $A$ is said to be *skew-symmetric* if and only if $A^t = -A$.

(i)  Prove that $A - A^t$ is skew symmetric for any three-by-three matrix $A$.

(ii)  Use Part (i) and the first result of Exercise 4 to show that any three-by-three matrix is the sum of a symmetric matrix and a skew symmetric matrix.

(iii)  Show that if a three-by-three matrix $A$ is either symmetric or skew symmetric, then $A^2$ is symmetric.

## 1.13   A Few Applications (*An Optional Section*)

To underscore the point that matrices and matrix algebra can be used to study a wide variety of problems, we shall examine briefly some applications in economics, sociology, and biology. Since real-world problems characteristically involve a large number of variables, we need to consider vectors with more than three components and matrices with more than three rows. In this way we shall lay the groundwork for a general study of linear algebra in subsequent chapters.

A real $m$-by-$n$ matrix $A$ is a rectangular array of real numbers arranged in $m$ rows and $n$ columns.

$$A = \begin{pmatrix} a_{11} & a_{12} & \cdots & a_{1n} \\ a_{21} & a_{22} & \cdots & a_{2n} \\ & \cdot & & \\ & \cdot & & \\ & \cdot & & \\ a_{m1} & a_{m2} & \cdots & a_{mn} \end{pmatrix}.$$

Since each row vector of $A$ has $n$ components, we can compute the scalar product of row $i$ of $A$ with any vector $X$ having $n$ components. If $X$ is written as a column vector (an $n$-by-one matrix),

$$X = \begin{pmatrix} x_1 \\ x_2 \\ \cdot \\ \cdot \\ \cdot \\ x_n \end{pmatrix},$$

then $AX$ can be defined to be the $m$-component column vector

$$AX = \begin{pmatrix} c_1 \\ c_2 \\ \cdot \\ \cdot \\ \cdot \\ c_m \end{pmatrix},$$

where

$$c_i = a_{i1}x_1 + a_{i2}x_2 + \ldots + a_{in}x_n \qquad \text{for } i = 1, 2, \ldots, m.$$

Similarly if $Y = (y_1 \ y_2 \ \ldots \ y_m)$ is an $m$-component row vector, then by forming the scalar product of $Y$ with each column vector of $A$ we obtain $n$ numbers that we write as a row vector:

$$YA = (b_1 \ b_2 \ \ldots \ b_n),$$

where

$$b_j = y_1 a_{1j} + y_2 a_{2j} + \ldots + y_m a_{mj} \qquad \text{for } j = 1, \ldots, n.$$

More generally a product of two rectangular matrices can be defined by means of the scalar product of vectors whenever the number of columns of the first matrix coincides with the number of rows of the second.

If $A = (a_{ij})$ is an $m$-by-$n$ matrix and if $B = (b_{ij})$ is an $n$-by-$p$ matrix, then $AB$ is an $m$-by-$p$ matrix,

$$AB = (c_{ij}),$$

where

$$c_{ij} = \sum_{k=1}^{n} a_{ik}b_{kj} \qquad \text{for } i = 1, \ldots, m \text{ and } j = 1, \ldots, p.$$

This definition clearly includes the cases we have seen previously; namely, $m = n = p = 3$; $m = 1, n = p = 3$; $m = n = 3, p = 1$; $m = 1, n$ and $p$ arbitrary; and $p = 1, m$ and $n$ arbitrary.

### An Example from Economics

A manufacturer requires a combination of various raw materials $M_1, \ldots, M_m$ to make several finished products, $P_1, \ldots, P_n$. Denote by $a_{ij}$ the amount of raw material $M_i$ needed to produce one unit of product $P_j$, and arrange this information in tabular form:

Products

| | | $P_1$ | $P_2$ | $\cdots$ | $P_n$ |

$$
\begin{array}{c|cccc}
 & P_1 & P_2 & \cdots & P_n \\
\hline
M_1 & a_{11} & a_{12} & \cdots & a_{1n} \\
M_2 & a_{21} & a_{22} & \cdots & a_{2n} \\
\cdot & \cdot & \cdot & & \cdot \\
\cdot & \cdot & \cdot & & \cdot \\
\cdot & \cdot & \cdot & & \cdot \\
M_m & a_{m1} & a_{m2} & \cdots & a_{mn}
\end{array}
$$

(Materials)

Now suppose that the monthly production schedule calls for an output of $u_j$ units of product $P_j$, where $j = 1, \ldots, n$. The amount of raw material $M_i$ needed to produce $u_j$ units of product $P_j$ is simply $a_{ij}u_j$, so the total monthly requirement of $M_i$ is

$$\sum_{j=1}^{n} a_{ij}u_j.$$

Thus if $U$ denotes the column vector describing the monthly production schedule, then the column vector $R$ that describes the monthly requirement of raw materials is given by

$$R = AU.$$

Similarly, let $p_i$ be the unit price of raw material $M_i$, and let $P$ denote the row vector $(p_1 \ p_2 \ \cdots \ p_m)$. Then the cost of the raw materials needed to produce one unit of product $P_j$ is

$$\sum_{i=1}^{m} p_i a_{ij},$$

and the total cost $C$ of raw materials for the monthly production schedule is given by

$$C = PR = PAU.$$

Clearly this scheme can be extended by including other economic factors, such as the unit requirements and costs of manpower and plant space for each product, and the costs of services and advertising. In a large economic model the size of the numbers $m$ and $n$ can be quite substantial.

A more complicated problem occurs if the available supplies of raw material are limited. Then we would seek to determine—among all monthly production schedules for which the requirements of raw materials do not exceed the restricted supplies—a schedule that maximizes some well-stated objective of the company, such as employment, profit, or production. Problems of this character are called *linear programming* problems, and their general solution uses techniques of linear algebra.

As an example, suppose that a farmer has two apple orchards, each producing apples in three grades of quality. He agrees to supply 12 bushels of Grade I, 8 of Grade II, and 24 of Grade III each day. The harvest in bushels and cost in dollars of each hour of picking are given by the table below.

|  | Grade I | Grade II | Grade III | Hourly cost |
|---|---|---|---|---|
| Orchard A | 6 | 2 | 4 | 20 |
| Orchard B | 2 | 2 | 12 | 16 |
| Daily requirements | 12 | 8 | 24 | |

The farmer desires to meet his commitment and to minimize the cost of picking. How many hours should he schedule for picking in each orchard?

If $x$ and $y$ denote the number of hours devoted to picking in Orchards A and B, respectively, then the constraints of the problem are

$$6x + 2y \geq 12,$$
$$2x + 2y \geq 8,$$
$$4x + 12y \geq 24,$$
$$x \geq 0,$$
$$y \geq 0,$$

and the problem is to choose $x$ and $y$ in order to

$$\text{minimize}\quad 20x + 16y$$

subject to the constraints. Geometrically the constraints determine a convex region $R$ of the $x$-$y$ plane, and for each number $c$, the equation

$$20x + 16y = c$$

is a line of slope $-\frac{5}{4}$. See Figure 1.22. Hence we seek the smallest value of $c$ such that the corresponding line intersects the region $R$. Geometrically we see that the minimum allowable value of $c$ occurs when

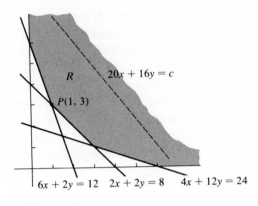

FIGURE 1.22

the line intersects $R$ only at the point $P(1, 3)$. Hence the farmer should schedule one hour of picking in Orchard A and three hours in Orchard B, at a minimal picking cost of 68 dollars.

Because this example involves only two variables, it can be solved geometrically by inspection. Most linear programs, however, contain many variables and require special algebraic methods. The very effective *simplex method* is based upon the pivot operation of Gauss-Jordan elimination.

### An Example from Sociology

In recent years attempts have been made to understand the mechanisms by which groups of people reach a consensus on an issue on which there is initially a spectrum of opinion. We imagine a small committee of persons $p_1, \ldots, p_n$, and attempt to describe the network of influence between pairs of individuals. A given person might be influenced by the opinions of some committee members but not others. The influence relationship within the committee can be recorded by means of a planar graph in which each person is represented by a point, and a directed line segment is drawn from $P_i$ to $P_j$ if the opinion of $p_i$ influences that of $p_j$. An example is sketched in Figure 1.23. Note that some lines are doubly directed, signifying mutual influence.

The same relationship can be shown by an $n$-by-$n$ matrix $A$ in which $a_{ij} = 1$ if $p_i$ influences $p_j$, and $a_{ij} = 0$ otherwise. Since we are interested in describing external influence we agree to let $a_{ii} = 0$. The incidence matrix for the graph of Figure 1.23 therefore is

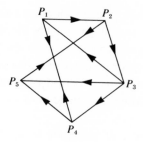

FIGURE 1.23

$$A = \begin{pmatrix} 0 & 1 & 0 & 1 & 0 \\ 0 & 0 & 1 & 0 & 1 \\ 1 & 0 & 0 & 1 & 1 \\ 1 & 0 & 0 & 0 & 1 \\ 0 & 1 & 0 & 0 & 0 \end{pmatrix}.$$

The sum $r_i$ of the entries in row $i$ is the number of persons influenced by $p_i$; equivalently, this is the number of directed line segments originating at $P_i$ in the graph. The sum $c_j$ of the entries in column $j$ is the number of persons who can influence $p_j$ (or the number of directed line segments terminating at $P_j$ in the graph). The numbers $r_i$ and $c_i$, therefore, are crude indicators of the influence and openmindedness, respectively, of $p_i$ relative to that committee.

Now consider the significance of $A^2$, where $A$ is an influence matrix; the element in row $i$ and column $j$ is

$$b_{ij} = a_{i1}a_{1j} + a_{i2}a_{2j} + \ldots + a_{in}a_{nj}.$$

A single term $a_{ik}a_{kj}$ will be 1 only when $p_i$ influences $p_k$ and $p_k$ influences $p_j$; otherwise it is 0. Hence $b_{ij}$ gives the number of persons through which $p_i$ can exert indirect (two-stage) influence on $p_j$. Equivalently, it is the number of distinct two-step directed paths on the graph from $P_i$ to $P_j$. For this example we have

$$A^2 = \begin{pmatrix} 1 & 0 & 1 & 0 & 2 \\ 1 & 1 & 0 & 1 & 1 \\ 1 & 2 & 0 & 1 & 1 \\ 0 & 2 & 0 & 1 & 0 \\ 0 & 0 & 1 & 0 & 1 \end{pmatrix}.$$

Observe that in this example each member can influence every other member in either one or two stages except for the pairs $(p_4, p_3)$, $(p_5, p_1)$, and $(p_5, p_4)$. As an exercise you may show that in this structure each member influences each other member in either one, two, or three stages. A "stage" may be regarded as a period of committee discussion, with opinions being recorded after each stage until a consensus is reached or a deadlock is evident.

Admittedly, this model of influence is far too simple to represent accurately the subtle intellectual and psychological factors involved in relationships between individuals. But the same model can be interpreted as describing a signaling or communications network wherein $p_i$ can send messages to some people and can receive messages from others. In that context we might ask whether it is possible for a given person $p_i$ to initiate a message (or start a rumor) that eventually will reach everyone, and if so, how many stages of transmission will be required.

### An Example from Biology

This example concerns population genetics and evolution in a species that reproduces sexually; that is, offspring are produced by a union of genetic material from two parents. For each genetic trait genetic information is contained in two types of genes, denoted G and g. Each individual carries a pair of those genes, so each individual can be classified by genotype GG, Gg, or gg with respect to that genetic trait. One gene of the pair is inherited from each parent, and the two genes of a parent are equally likely to be transmitted to the offspring. For example, the offspring of parents of genotype GG and Gg must inherit a G gene from the former; the gene inherited from the latter will be G with probability $\frac{1}{2}$. Hence the genotype of the offspring will be GG with probability $\frac{1}{2}$, Gg with probability $\frac{1}{2}$, and gg with probability 0.

The probabilities of offspring resulting from matings of all possible genotypes can be displayed in a table, as shown on the next page.

Note that each *row vector* of the three matrices $P$, $Q$, and $R$ has two very special properties.

Each component $x$ satisfies $0 \le x \le 1$.
The sum of the components is 1.

| Male parent | Female parent | Probability of offspring of type | | |
|:---:|:---:|:---:|:---:|:---:|
| | | GG | Gg | gg |
| GG | GG | $1$ | $0$ | $0$ |
| GG | Gg | $\frac{1}{2}$ | $\frac{1}{2}$ | $0$ |
| GG | gg | $0$ | $1$ | $0$ |
| Gg | GG | $\frac{1}{2}$ | $\frac{1}{2}$ | $0$ |
| Gg | Gg | $\frac{1}{4}$ | $\frac{1}{2}$ | $\frac{1}{4}$ |
| Gg | gg | $0$ | $\frac{1}{2}$ | $\frac{1}{2}$ |
| gg | GG | $0$ | $1$ | $0$ |
| gg | Gg | $0$ | $\frac{1}{2}$ | $\frac{1}{2}$ |
| gg | gg | $0$ | $0$ | $1$ |

The first three offspring-probability rows form the matrix $= P$, the next three $= Q$, and the last three $= R$.

A vector having these two properties is called a *probability vector*, and a square matrix each of whose rows is a probability vector is called a *Markov* (or *stochastic*) *matrix*. It is not difficult to prove that the product of two Markov matrices is a Markov matrix, and that any *convex* linear combination of Markov matrices,

$$a_1 M_1 + a_2 M_2 + \ldots + a_k M_k, \qquad \text{where } 0 \leq a_i \leq 1 \text{ and } \sum_{i=1}^{k} a_i = 1,$$

is a Markov matrix.

We now make the following three assumptions:

The population under study is large.

Mating is random with respect to genotype.

The distribution of genotypes within the population is independent of sex and life expectancy.

Let the probability vector $D_0 = (p, q, r)$ denote the proportions of the adult population of genotypes GG, Gg, and gg, respectively, at the start of the experiment. The third assumption means that the distributions of genotypes among adult males and among adult females can also be described by $D_0$; moreover, for any generation of offspring with a distribution of genotypes described by a probability vector $D_1$, the proportions that survive to breed are also described by $D_1$. Change in the distribution vectors $D_1, D_2, D_3, \ldots$ of successive generations indicates that evolution is taking place within the population.

Let us digress for a moment to illustrate the special case of a controlled experiment in which the only males that are permitted to breed are of genotypes GG, but these breed randomly. In a large number of matings of GG males with females of genotype distribution $D = (p, q, r)$, all of a proportion $p$ of the matings will produce GG offspring, one half of a proportion $q$ of the matings will produce GG offspring, and none of a proportion $r$ of the matings will produce GG offspring. If we make a similar analysis for the production of Gg and gg offspring, we note that the distribution of genotypes in the first generation can be described by the vector

$$D_1 = D_0P = (p, q, r) \begin{pmatrix} 1 & 0 & 0 \\ \frac{1}{2} & \frac{1}{2} & 0 \\ 0 & 1 & 0 \end{pmatrix} = (p + \frac{q}{2}, \frac{q}{2} + r, 0).$$

If the experiment permitted only Gg males to breed, the matrix $Q$ would be used in place of $P$, and so on, in the table on page 77.

Now consider the case where all males are permitted to breed randomly. To describe this experiment we form a matrix $M$ as a convex linear combination of the matrices $P$, $Q$, and $R$, weighted by the corresponding proportions in which each would be found in a large number of matings:

$$M = pP + qQ + rR = \begin{pmatrix} p + \frac{q}{2} & \frac{q}{2} + r & 0 \\ \frac{p}{2} + \frac{q}{4} & \frac{p}{2} + \frac{q}{2} + \frac{r}{2} & \frac{q}{4} + \frac{r}{2} \\ 0 & p + \frac{q}{2} & \frac{q}{2} + r \end{pmatrix}.$$

To write $M$ in simpler form we let

$$a = p + \frac{q}{2},$$

$$b = \frac{q}{2} + r,$$

and note that $a + b = p + q + r = 1$. (Biologically $a$ represents the proportion of G genes in the population, and $b$ represents the proportion of g genes.) Hence

$$M = \begin{pmatrix} a & b & 0 \\ \dfrac{a}{2} & \dfrac{1}{2} & \dfrac{b}{2} \\ 0 & a & b \end{pmatrix}.$$

Because of the relations between $a$, $b$, $p$, $q$, and $r$, the result of the computation $D_0 M$ can be reduced to the form

$$D_1 = D_0 M = (a^2, 2ab, b^2),$$

which describes the genotype distribution of the first generation of offspring.

To see the effect of unrestricted matings among the first generation we must first calculate a new transition matrix $M_1$ using the genotype distribution $(a^2, 2ab, b^2)$ in place of $(p, q, r)$. $M_1$ can be obtained from $M$ by making these substitutions:

$$p^2 + \frac{q}{2} = a^2 + ab = a(a + b) = a,$$

$$\frac{q}{2} + r = ab + b^2 = b(a + b) = b.$$

It is interesting that we obtain $M_1 = M$, and the result of computing the genotype distribution of the offspring of the first generation is

$$D_2 = D_1 M_1 = D_1 M = (a^2, 2ab, b^2) = D_1.$$

Hence $D_1$ is a fixed point of $M$, a characteristic vector associated with the characteristic value 1.

The genetic interpretation of these results can now be summarized. We begin with a population in which the proportion of genotypes GG, Gg, and gg are

$$D_0 = (p, q, r).$$

The distribution of genotypes in the first generation of offspring is

$$D_1 = \left( \left( p + \frac{q}{2} \right)^2, 2\left( p + \frac{q}{2} \right)\left( \frac{q}{2} + r \right), \left( \frac{q}{2} + r \right)^2 \right),$$

which in general is different from $D_0$. However, the offspring of the first and all succeeding generations have the same genotype distribution, $D_1$. In short, genetic equilibrium is achieved in a single generation under the conditions assumed for this model.

This result, known as the Hardy-Weinberg law, was announced in 1908 independently by both the noted English mathematician G. H. Hardy and the German physician W. Weinberg.

We make one further observation: the symmetry of the model for genotypes GG and gg make it reasonable to investigate the special case in which $p = r$. Then $2p + q = 1$, so the distribution vector at genetic equilibrium is

$$D_1 = \left(\frac{1}{4}, \frac{1}{2}, \frac{1}{4}\right).$$

This distribution of probabilities occurs frequently in the study of population genetics.

EXERCISES 1.13

1. Graph each of the following linear inequalities and shade the region $R$ containing all points $(x, y)$ that satisfy all of the inequalities simultaneously.

$$2x + \phantom{8}y \geq 12,$$
$$5x + 8y \geq 74,$$
$$x + 6y \geq 24,$$
$$x \geq 0,$$
$$y \geq 0.$$

2. Referring to Exercise 1, for each of the following linear forms find a point of $R$ at which that linear form attains its minimum value on $R$.

(i)  $x + y.$

(ii)  $3x + y.$

(iii)  $x + 3y.$

(iv)  $3x - y.$

3. Given the inequalities

$$x + 2y \leq 720,$$
$$5x + 4y \leq 1800,$$
$$3x + \phantom{4}y \leq 900,$$
$$x \geq 0,$$
$$y \geq 0.$$

Sketch the region $R$ of simultaneous solutions, and for each of the following linear forms determine a point of $R$ at which that linear form attains its maximum value on $R$.

(i) $40x + 50y$.

(ii) $40x + 70y$.

(iii) $40x + 20y$.

4. Refer to the text example of a graph and its incidence matrix $A$.

(i) Compute $A^2$ to verify the results stated.

(ii) Compute $A + A^2 + A^3$ to verify that each entry is positive.

5. Write the six-by-six incidence matrix $B$ for the directed graph sketched below, compute $B^2$ to determine the two-step relationships, and draw a graph that corresponds to $B^2$ (a graph of the two-step channels.)

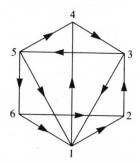

6. In an influence (or communications) matrix, what is the significance of a row in which every entry is zero? A column in which each entry is zero?

7. Prove that if $A$ and $B$ are any three-by-three Markov matrices and if $a$ and $b$ are nonnegative numbers such that $a + b = 1$, then

(i) $AB$ is a Markov matrix,

(ii) $aA + bB$ is a Markov matrix.

Would your argument be valid for $n$-by-$n$ Markov matrices?

8. Refer to the text example concerning genetics. Verify by matrix computations the assertions that

(i) $D_1 = (a^2, 2ab, b^2)$, where $D_1 = D_0 M$,

(ii) $M_1 = M$, where $M_1 = a^2 P + 2abQ + b^2 R$,

(iii) $D_2 = D_1$, where $D_2 = D_1 M_1$.

# 2

# *Vector Spaces*

Having acquired some background information about linear algebra in the special context of $R_3$, we now need to develop a broader setting in which the major concepts of linear algebra can be studied more precisely and in greater detail. A brief enumeration of these ideas serves both to summarize the content of Chapter 1 and to indicate the emphasis of succeeding chapters.

*Vector Spaces.* The underlying structure of linear algebra is a vector space; related notions are subspaces, linear independence, bases, inner products, and the geometric significance of all of these.

*Matrices.* Matrices can be used to represent various mathematical entities; special emphasis is given to systems of linear equations, linear transformations, and quadratic forms. Algebraic operations on matrices are chosen so as to provide algebraic means of investigating the properties of the objects thus represented.

*Systems of Linear Equations.* One method of solving a given system of linear equations is to replace it by another system having precisely the same solutions but written in such a form that its solution is readily obtained. Matrix operations provide a convenient scheme for performing such replacements.

*Linear Transformations.*    The matrix representation of a linear transformation depends upon the choice of basis (coordinate system). Geometric properties of the transformation are reflected as algebraic properties of the corresponding matrices. Again, a central problem is to obtain the simplest possible matrix representation. Likewise, the representation of a quadratic form by a matrix depends upon a choice of basis for the underlying vector space, and geometric properties of the form can be interpreted as numerical properties of the corresponding matrices. Simple representations, therefore, are of special interest.

## 2.1   Vector Spaces and Subspaces

A vector space is an algebraic system based upon two sets $F$ and $V$ of elements and various ways of combining those elements to produce other elements of the system. The elements of $F$ are called *scalars*, and those of $V$ are called *vectors*.

Scalars may be regarded as numbers; for most of our study we shall take $F$ to be the set $R$ of real numbers, but sometimes it will be convenient to choose $F$ to be the set $C$ of complex numbers. Both $R$ and $C$ are examples of a general algebraic system that is called a *field*. If you aren't familiar with the definition of a field, you can find it in almost any introductory book on modern (or abstract) algebra. However, except for a few special topics, the particular choice of $F$ will be immaterial in our study.

Vectors, on the other hand, can be any kind of object for which suitable notions of addition and scalar multiple are defined. By this we mean that the "sum" of any two vectors of $V$ is a uniquely determined vector of $V$, that the "multiple" of a scalar of $F$ and a vector of $V$ is a vector of $V$, and that these operations satisfy the properties listed formally below.

**DEFINITION 2.1**    *A vector space $\mathcal{V}$ over a field $F$ is an algebraic system $\mathcal{V}(F) = \{V, F; + , \cdot , \oplus , \odot\}$ that satisfies the following postulates:*

(a)   $\{F; +, \cdot\}$ is a field;

(b)   For all $\alpha, \beta \in V$, $\alpha \oplus \beta \in V$, and

  (1)   $\alpha \oplus \beta = \beta \oplus \alpha$,

  (2)   $(\alpha \oplus \beta) \oplus \gamma = \alpha \oplus (\beta \oplus \gamma)$      for all $\gamma \in V$,

  (3)   there exists $\theta \in V$ such that $\alpha \oplus \theta = \alpha$,

  (4)   for each $\alpha \in V$, there exists $\alpha' \in V$ such that $\alpha \oplus \alpha' = \theta$;

(c)   For all $a, b \in F$ and all $\alpha, \beta \in V$, $a \odot \alpha \in V$, and

  (5)   $(a + b) \odot \alpha = (a \odot \alpha) \oplus (b \odot \alpha)$,

  (6)   $a \odot (\alpha \oplus \beta) = (a \odot \alpha) \oplus (a \odot \beta)$,

  (7)   $(a \cdot b) \odot \alpha = a \odot (b \odot \alpha)$,

  (8)   $1 \odot \alpha = \alpha$.

The vector $\theta$ is called the *zero vector*, and the vector $\alpha'$ is called the *negative* of $\alpha$, denoted $-\alpha$.

As we observed for $R_3$, the use of lower case Latin letters for scalars and lower case Greek letters for vectors removes any possible ambiguity about which kind of sum and which kind of product is intended, so we now replace the symbol $\oplus$ by $+$ and omit the symbols $\cdot$ and $\odot$ entirely. Also we frequently write $\mathscr{V}$ instead of $\mathscr{V}(F)$.

**THEOREM 2.1**   In any vector space

(a)   $\theta$ is unique,

(b)   $-\alpha$ is uniquely determined by $\alpha$,

(c)   $0\alpha = \theta$,

(d)   $(-1)\alpha = -\alpha$,

(e)   $k\theta = \theta$ for every scalar $k$.

PROOF   The proofs of these basic results are not difficult, but to illustrate the methods we shall prove (a), (c), and (d) in detail and ask you to prove (b) and (e) as exercises.

  To prove that there is only one zero vector, we suppose there are two, say $\theta$ and $\omega$, both satisfying Postulate (3) and prove that they must be equal. We have

$$\omega = \omega + \theta \quad \text{by (3)}$$
$$= \theta + \omega \quad \text{by (1)}$$
$$= \theta,$$

where in the last step we use the assumption that $\omega$ satisfies $\alpha + \omega = \alpha$ for all $\alpha \in V$.

To prove (c) we write

$$\alpha = 1\alpha \quad\quad \text{by (8)}$$
$$= (0 + 1)\alpha$$
$$= 0\alpha + 1\alpha \quad \text{by (5)}$$
$$= 0\alpha + \alpha \quad \text{by (8)}.$$

Thus $\alpha + (-\alpha) = (0\alpha + \alpha) + (-\alpha) = 0\alpha + (\alpha + (-\alpha)) \quad$ by (2);

$$\theta = 0\alpha + \theta \quad \text{by (4)}$$
$$= 0\alpha \quad\quad \text{by (3)}.$$

To prove (d) we use the result (b), which can be proved directly from the postulates of a vector space.

$$\alpha + (-1)\alpha = 1\alpha + (-1)\alpha \quad\quad \text{by (8)}$$
$$= (1 + (-1))\alpha \quad\quad \text{by (5)}$$
$$= 0\alpha = \theta \quad\quad\quad \text{by (c)}.$$

But by (b) there is only one vector, $-\alpha$, that satisfies

$$\alpha + (-\alpha) = \theta,$$

so

$$(-1)\alpha = -\alpha.$$

### Examples of Vector Spaces

(a)   For any positive integer $n$, the set of all ordered $n$-tuples $(x_1, x_2, \ldots, x_n)$ of real numbers, with vector sum and scalar multiple defined component-by-component, is called Cartesian $n$-space, $R_n$. $R_1$

is simply the field $R$ of real numbers; $R_2$ is the real plane; and $R_3$ is Cartesian 3-space.

(b)   A trivial example is the *zero space*, which contains only one vector, $\theta$.

(c)   Let $F$ be any field; we consider the elements of $F$ both as vectors and as scalars, with vector sum and scalar multiple defined as in $F$. In Section 1.1 $R$ was our first example of a vector space.

(d)   Let $V$ be the set of all polynomials with real coefficients and of degree not exceeding $k$. A vector thus is a function of the form

$$p(x) = a_0 + a_1 x + \ldots + a_k x^k,$$

and the sum and scalar multiples of vectors are defined as for polynomials. It is easy to discover that this space $\mathscr{P}_k$ is a replica of $R_{k+1}$.

(e)   The space $\mathscr{P}$ of all real polynomials is defined as for $\mathscr{P}_k$, except that polynomials of *all degrees* constitute the set of vectors.

(f)   The set of all real valued functions that are continuous on the closed interval $a \le x \le b$ forms a vector space $\mathscr{C}[a, b]$, with the operations defined by

$$(f \oplus g)(x) = f(x) + g(x),$$

$$(k \odot f)(x) = kf(x).$$

(g)   The set of all solutions of the differential equation

$$y'' + p(x)y' + q(x)y = 0$$

also forms a vector space of functions.

(h)   The set of all solutions $(x, y, z)$ of the system of linear equations

$$a_1 x + a_2 y + a_3 z = 0,$$

$$b_1 x + b_2 y + b_3 z = 0,$$

$$c_1 x + c_2 y + c_3 z = 0,$$

forms a vector space, where the sum and scalar multiple of triples are defined as for $R_3$.

You should take time to verify that each of these examples satisfies all the requirements of a vector space as specified in Definition 2.1. In particular, you must be sure that the given set of "vectors" is *closed under the given operations:* $\alpha + \beta \in V$ and $c\alpha \in V$ for all $\alpha$, $\beta \in V$, $c \in F$.

A moment's reflection will reveal a close relationship between certain of these examples. The vectors of Example (d) form a subset of those of Example (e); and those of (h) are a subset of $R_3$. Moreover, the single vector of Example (b) is a subset of the vectors of any vector space. Each of these illustrates the concept of a *subspace*.

> **DEFINITION 2.2**    A nonvoid subset $S$ of vectors of a vector space $\mathscr{V}(F)$ is said to form a *subspace* of $\mathscr{V}$ if and only if the subsystem $\mathscr{S} = \{S, F; +, \cdot, \oplus, \odot\}$ is a vector space.

By the definitions of vector space and subspace, if a subset $S$ forms a subspace of $\mathscr{V}$, the operations involving vectors must be closed operations on the set $S$; that is

$$\alpha + \beta \in S \qquad \text{for each } \alpha, \beta \in S,$$

$$c \cdot \alpha \in S \qquad \text{for each } \alpha \in S \text{ and each } c \in F.$$

It is easy to prove that this condition is sufficient also. For then $(-1)\alpha = -\alpha \in S$, and hence $\alpha + (-\alpha) = \theta \in S$. The associativity and commutativity of vector addition and the required properties of scalar multiplication are inherited in $S$ from the corresponding universal properties of $\mathscr{V}$. Hence $S$ satisfies the axioms of a vector space. We have proved:

> **THEOREM 2.2**    A nonvoid subset of a vector space is a subspace if and only if the subset is closed with respect to vector sum and scalar multiple.

### Further Examples of Subspaces

(a)    In $R_3$, we have already seen that the subspaces are the zero vector itself, any line through $\theta$, any plane through $\theta$, or $R_3$ itself. There are no others.

(b)    In the space of all polynomials of degree $\leq n$, for any $k \leq n$ the set of all polynomials of degree $\leq k$ is a subspace.

(c)    In the space of all real functions continuous on the interval $[0, 1]$, the set of all functions that are constant on $[0, 1]$ is a subspace.

(d)   In the space $R_n$ of all real $n$-tuples, the set of all $n$-tuples for which the first entry is 0 is a subspace. Likewise the set of all $n$-tuples $(x_1, \ldots, x_n)$ for which $x_n = \sum_{i=1}^{n-1} x_i$ is a subspace.

Our next example is more general. Let $S$ be a nonvoid subset of vectors of an arbitrary vector space $\mathscr{V}$. Denote by $[S]$ the set of *all* linear combinations (finite sums of scalar multiples) of vectors of $S$. Then by Theorem 2.2, $[S]$ is a subspace of $\mathscr{V}$ called the subspace *spanned* by (or *generated* by) $S$.

Now let $\mathscr{S}$ and $\mathscr{T}$ be subspaces of $\mathscr{V}$. The set intersection $S \cap T$ is easily proved to be a subspace also. But the set union $S \cup T$ generally is *not* a subspace, because a subspace containing $S$ and $T$ must also contain all linear combinations of vectors of $S$ and vectors of $T$.

Hence the *sum* $\mathscr{S} + \mathscr{T}$ of two subspaces of $\mathscr{V}$ is defined to be the space $[S \cup T]$ spanned by the set union of the vectors of $\mathscr{S}$ and the vectors of $\mathscr{T}$.

**DEFINITION 2.3**

(a)   If $S$ is a nonvoid subset of vectors of the vector space $\mathscr{V}$, the set $[S]$ of all linear combinations of vectors of $S$ is called the *subspace of $\mathscr{V}$ spanned by $S$.*

(b)   The *sum* $\mathscr{S} + \mathscr{T}$ of two subspaces of $\mathscr{V}$ is the subspace $[S \cup T]$ spanned by the set union of the vectors of $\mathscr{S}$ and the vectors of $\mathscr{T}$.

**THEOREM 2.3**   If $\mathscr{S}$ and $\mathscr{T}$ are subspaces of $\mathscr{V}$, then

$$\mathscr{S} + \mathscr{T} = \{\sigma + \tau \mid \sigma \in S \text{ and } \tau \in T\}.$$

**PROOF**   By definition $\mathscr{S} + \mathscr{T}$ is the set of *all* linear combinations of vectors of $S$ and of $T$ and therefore $\mathscr{S} + \mathscr{T}$ certainly contains all vectors of the form $\sigma + \tau$. But given any linear combination of vectors of $S \cup T$, we can first write the vectors in $S$ and then in $T$ since vector addition is commutative and associative: we then have

$$(c_1\sigma_1 + \ldots + c_r\sigma_r) + (d_1\tau_1 + \ldots + d_s\tau_s),$$

which is of the form $\sigma + \tau$ since the first linear combination of vectors of $S$ must be a vector of $\mathscr{S}$, because $\mathscr{S}$ is a subspace. Similarly the second expression in parentheses is a vector of $\mathscr{T}$.

Since $\theta$ is in any subspace, the smallest possible intersection that any two subspaces can have is the subspace $[\theta]$. Frequently we are interested in decomposing $\mathscr{V}$ into the sum of two subspaces as efficiently as possible; that is, one having the smallest possible overlap. We describe that situation by a special term.

**DEFINITION 2.4**    A vector space $\mathscr{V}$ is said to be the *direct sum* of two subspaces $\mathscr{S}$ and $\mathscr{T}$ if and only if

$$\mathscr{V} = \mathscr{S} + \mathscr{T},$$

$$[\theta] = \mathscr{S} \cap \mathscr{T}.$$

When this is so we write $\mathscr{V} = \mathscr{S} \oplus \mathscr{T}$.

**THEOREM 2.4**    $\mathscr{V} = \mathscr{S} \oplus \mathscr{T}$ if and only if each $\xi \in \mathscr{V}$ has a *unique* representation $\xi = \sigma + \tau$ for some $\sigma \in \mathscr{S}$ and some $\tau \in \mathscr{T}$.

**PROOF**    Let $\mathscr{V} = \mathscr{S} \oplus \mathscr{T}$. Then $\mathscr{V} = \mathscr{S} + \mathscr{T}$ so that each $\xi \in \mathscr{V}$ can be written as the sum $\xi = \sigma + \tau$ for suitable $\sigma \in \mathscr{S}, \tau \in \mathscr{T}$. Suppose there were two such representations of $\xi$,

$$\xi = \sigma_1 + \tau_1,$$

$$\xi = \sigma_2 + \tau_2.$$

Then

$$\theta = (\sigma_1 - \sigma_2) + (\tau_1 - \tau_2),$$

$$\sigma_2 - \sigma_1 = \tau_1 - \tau_2.$$

Since $\mathscr{S}$ and $\mathscr{T}$ are subspaces, the vectors $\sigma_2 - \sigma_1$ and $\tau_1 - \tau_2$ are in $\mathscr{S}$ and in $\mathscr{T}$, respectively. But since they are equal, they are in $\mathscr{S} \cap \mathscr{T} = [\theta]$. Hence $\sigma_2 - \sigma_1 = \theta = \tau_1 - \tau_2$, so $\sigma_2 = \sigma_1$ and $\tau_1 = \tau_2$.

Conversely, if each vector $\xi \in \mathscr{V}$ has a *unique* representation $\xi = \sigma + \tau$ then clearly $\mathscr{V} = \mathscr{S} + \mathscr{T}$. We need only to prove that $\mathscr{S} \cap \mathscr{T} = [\theta]$. Suppose that $\alpha \in \mathscr{S} \cap \mathscr{T}$. Then we can write

$$\alpha = \alpha + \theta, \quad \text{where } \alpha \in \mathscr{S} \text{ and } \theta \in \mathscr{T},$$

$$\alpha = \theta + \alpha, \quad \text{where } \theta \in \mathscr{S} \text{ and } \alpha \in \mathscr{T}.$$

Since any representation of this form is unique, we have $\alpha = \theta$, and hence $\mathscr{S} \cap \mathscr{T} = [\theta]$.

In the proof of Theorem 2.4 you will observe a convenient misuse of notation. Heretofore we have carefully distinguished between a vector space $\mathscr{V}$ and the set $V$ of vectors of that system. Inasmuch as there is no possibility of ambiguity we shall write $\alpha \in \mathscr{V}$ when what we really mean is $\alpha \in V$.

## EXERCISES 2.1

1. Let $\mathscr{S}$ denote the set of all infinite sequences of real numbers, with the sum and scalar multiples of sequences defined term by term, as usual. Let $\mathscr{K}$ denote the set of all convergent sequences. Verify that $\mathscr{S}$ is a real vector space and that $\mathscr{K}$ is a subspace of $\mathscr{S}$.

2. Show that the set of all solutions of the differential equation

$$y'' + p(x)y' + q(x)y = 0$$

form a real vector space. (You do not need to know the specific functions that are solutions in order to show that they form a vector space.)

3. Explain why the polynomial space $\mathscr{P}_k$ can be regarded as a replica of $R_{k+1}$.

4. Prove Part (b) of Theorem 2.1.

5. Prove Part (e) of Theorem 2.1.

6. Which of the following subsets of $R_n$ are *not* subspaces? Cite a specific reason in each case:

   (i)   $\{(a_1, \ldots, a_n) \mid a_1 < a_2\}$.

   (ii)  $\{(a_1, \ldots, a_n) \mid a_1 + \ldots + a_n = 0\}$.

   (iii) $\{(a_1, \ldots, a_n) \mid a_1 + \ldots + a_n = 1\}$.

   (iv) $\{(a_1, \ldots, a_n) \mid a_1 = 2a_2 + \ldots + na_n\}$.

   (v)  $\{(a_1, \ldots, a_n) \mid a_1 = 0 = a_n\}$.

7. Let $\mathscr{C}[0, 1]$ be the space of all real functions that are continuous on $[0, 1]$. Which of the following subsets are subspaces? Explain your answers.

   (i)   All polynomials of degree $\geq 2$.

   (ii)  All polynomials of degree 2.

   (iii) All odd functions: $f(-x) = -f(x)$ for all $x$.

   (iv) All functions such that $|f(x)| \leq 0$ for all $x$.

8. Let $\mathscr{V}$ denote the vector space of all real functions defined on the interval $[-1, 1]$. Determine whether or not each of the following subsets of $\mathscr{V}$ is a subspace.

(i) $S_1 = \{f \in \mathscr{V} \mid f(0) = f(1)\}$.

(ii) $S_2 = \{f \in \mathscr{V} \mid f(0) = 1\}$.

(iii) $S_3 = \{f \in \mathscr{V} \mid f(x^2) = f^2(x)\}$.

(iv) $S_4 = \{f \in \mathscr{V} \mid f(x) = a \sin(x + b)$ for some $a$ and $b \in R\}$.

9. Let $\mathscr{C}[-1, 1]$ be the space of all real functions that are continuous on $[-1, 1]$. Which of the following subsets of $\mathscr{C}[-1, 1]$ are subspaces? Explain your answers.

(i) All $f$ that intersect the $x$-axis at $x = 1$.

(ii) All $f$ that have a point of inflection at $x = 0$.

(iii) All $f$ that have either a relative maximum or a relative minimum at $x = 0$.

10. Referring to Exercise 1 of Section 1.3, let $\mathscr{S} = [\alpha, \beta]$ and $\mathscr{T} = [\gamma, \delta]$. Describe geometrically $\mathscr{S} \cap \mathscr{T}$ and $\mathscr{S} + \mathscr{T}$ for each of the examples listed.

11. Let $\mathscr{S}, \mathscr{T}, \mathscr{U}$ be subspaces of $\mathscr{V}$.

(i) Prove that $(\mathscr{S} + \mathscr{T}) \cap \mathscr{U} \supseteq (\mathscr{S} \cap \mathscr{U}) + (\mathscr{T} \cap \mathscr{U})$.

(ii) Show by an example in the plane that equality need not hold.

(iii) Prove that if $\mathscr{S} \subseteq \mathscr{U}$, equality does hold.

12. Show that the set of all real valued functions that have a derivative at each point of the interval $a \leq x \leq b$ forms a vector space. Is this space a subspace of the vector space described in Example (f)?

## 2.2   Linear Independence

We define linear dependence and independence in arbitrary vector spaces exactly as we did for $R_3$.

**DEFINITION 2.5**   Let $\mathscr{V}$ be a vector space. A set $S$ of vectors of $\mathscr{V}$ is said to be *linearly dependent* if and only if there exist a finite set $\{\alpha_1, \ldots, \alpha_k\}$ of vectors of $S$ and a set of scalars $\{c_1, \ldots, c_k\}$ *not all of which are zero* such that

$$c_1\alpha_1 + c_2\alpha_2 + \ldots + c_k\alpha_k = \theta.$$

Otherwise $S$ is said to be *linearly independent*.

A consequence of this definition is that the void set is linearly independent. We arbitrarily define the space spanned by the void set to be the zero subspace $[\Phi] = [\theta]$. Previously $[S]$ had been defined only for a non-void set $S$, so this is an extension of Definition 2.3. Not only is it logically valid to extend the definition in this way but it is convenient not to have to make an exceptional case of the void set.

The results of Exercise 4 of Section 1.5 remain valid in any vector space:

**THEOREM 2.5**   Any subset of a linearly independent set is linearly independent, and any set containing a linearly dependent set is itself linearly dependent.

To construct linearly independent sets in $\mathscr{V}$ we can proceed inductively. Beginning with any linearly independent set $S$ (for example, the void set), we can extend $S$ to a larger linearly independent set by adjoining any vector that does not already lie in $[S]$. Such a vector will exist unless $[S] = \mathscr{V}$.

**THEOREM 2.6**   Let $S$ be a linearly independent set of vectors of $\mathscr{V}$. For any vector $\xi \in \mathscr{V}$, the set $S \cup \{\xi\}$ is linearly independent if and only if $\xi \notin [S]$.

**PROOF**   If $\xi \in [S]$, either $\xi = \theta$ or $\xi$ is a nontrivial linear combination of vectors of $S$, so $S \cup \{\xi\}$ is linearly dependent. Conversely, if $\xi \notin [S]$ consider any equation of the form

$$a_1\sigma_1 + \ldots + a_k\sigma_k + b\xi = \theta,$$

where $\sigma_i \in S$ for each $i$. If $b \neq 0$ then $\xi \in [S]$ contrary to hypothesis. Hence $b = 0$. Then also each $a_i = 0$ since the $\sigma_i$ form a linearly independent set. Hence $S \cup \{\xi\}$ is linearly independent.

**THEOREM 2.7**   If $S$ is a finite set of vectors that spans a subspace $\mathscr{S}$, then a linearly independent subset of $S$ also spans $\mathscr{S}$.

**PROOF**   If $\mathscr{S} = [\theta]$, the void set is independent and by definition spans $\mathscr{S}$. Otherwise, $S$ is nonvoid; choose $\sigma_1 \neq \theta$ from $S$. Then $\{\sigma_1\}$ is independent. If $[\sigma_1]$ is a proper subspace of $\mathscr{S}$, by Theorem 2.6 there exists $\sigma_2 \in S$ such that $\{\sigma_1, \sigma_2\}$ is independent. This argument can be repeated until $[\sigma_1, \ldots, \sigma_k] = \mathscr{S}$.

Observe that Theorem 2.6 confirms our earlier observation about $R_3$—a linearly independent set $S$ can be extended to a larger independent set by adjoining any vector that is *not* in $[S]$. Actually Theorem 2.7 remains valid if the word "finite" is deleted, but a more general form of proof is then needed, and we shall not pursue that idea.

Notice that the proof of Theorem 2.7 started with a small independent set, which was successively enlarged to obtain an independent set that spans $\mathscr{S}$. It is instructive to proceed in the other direction, starting with a large dependent set and successively removing vectors that make no essential contribution in spanning $\mathscr{S}$. We illustrate the process in proving the following theorem, of which Theorem 2.7 is an easy consequence.

**THEOREM 2.8**  Let $S$ be a finite, linearly dependent set, and let the vectors of $S$ be written in any order, $S = \{\sigma_1, \ldots, \sigma_m\}$. Some vector $\sigma_p$ is a linear combination of vectors that precede $\sigma_p$ in this order.
**PROOF**  Since $S$ is linearly dependent, there must exist scalars $c_1, \ldots, c_m$, not all zero, such that

$$c_1\sigma_1 + \ldots + c_m\sigma_m = \theta.$$

Let $c_p$ be the last nonzero scalar coefficient in this equation. Then

$$\sigma_p = -c_p^{-1}(c_1\sigma_1 + \ldots + c_{p-1}\sigma_{p-1}),$$

which completes the proof. We observe further that $[S_1] = [S]$, where $S_1$ denotes the set $S$ with $\sigma_p$ *removed*.

The construction of a linearly independent spanning set for a subspace $\mathscr{S}$ can be accomplished, therefore, either by starting with a small linearly independent subset and enlarging it, retaining independence until it spans $\mathscr{S}$, or by starting with a large spanning set for $\mathscr{S}$ and reducing it, retaining a spanning set for $\mathscr{S}$ until it is linearly independent. In the previous chapter a linearly independent spanning set was called a basis. We shall resume the study of bases in the next section.

When the vectors of $\mathscr{V}$ can be represented as $n$-tuples of numbers, Theorem 2.8 suggests a computational method for establishing linear dependence or independence of a given set of vectors, $S = \{\sigma_1, \ldots, \sigma_m\}$, where $\sigma_i = (s_{i1}, \ldots, s_{in})$. Consider the $m$-by-$n$ matrix $M$ in which row $i$ is $\sigma_i$, $i = 1, \ldots, m$,

$$M = \begin{pmatrix} s_{11} & s_{12} & \cdots & s_{1n} \\ s_{21} & s_{22} & \cdots & s_{2n} \\ & \cdot & & \cdot \\ & \cdot & & \cdot \\ & \cdot & & \cdot \\ s_{m1} & s_{m2} & \cdots & s_{mn} \end{pmatrix}.$$

According to Theorem 2.8, if $S$ is dependent then some row is a linear combination of previous rows. Then a suitable sequence of elementary row operations (described for three-by-three matrices in Section 1.4) will replace $M$ by a matrix having at least one complete row of zeros. On the other hand if $S$ is linearly independent, a suitable sequence of elementary row operations will replace $M$ by a matrix having row echelon form and $m$ nonzero rows.

In summary: to test linear independence of $m$ given $n$-tuples, we can form the $m$-by-$n$ matrix $M$ having those $n$-tuples as row vectors. We then use elementary row operations to reduce $M$ to echelon form in which the first nonzero entry in each nonzero row appears in a later column than does the first nonzero entry of any previous row. The original set is linearly independent if and only if no zero row occurs in the echelon form. More generally, the number of nonzero rows of the echelon form is the number of vectors in a largest linearly independent subset of the given $n$-tuples.

## EXERCISES 2.2

1. Select a maximal linearly independent subset of each of the following sets of vectors.

(i)  (1, 0, 1, 0), ( 0, 1, 0, 1), (1, 1, 1, 1), (−1, 0, 2, 0).

(ii)  (0, 1, 2, 3), ( 3, 0, 1, 2), (2, 3, 0, 1), ( 1, 2, 3, 0).

(iii) (1, −1, 1, −1), (−1, 1, −1, −1), (1, −1, 1, −2), ( 0, 0, 0, 1).

2. For each example of the preceding exercise in which the largest independent subset contains fewer than four vectors, adjoin vectors $(a_1, a_2, a_3, a_4)$ to obtain a linearly independent set of four vectors.

3. If $\{\alpha, \beta, \gamma\}$ is linearly independent, determine whether or not each of the following sets is linearly independent:

(i)  $\{\alpha + \beta, \beta + \gamma, \gamma + \alpha\}$;

(ii)  $\{\alpha - \beta, \beta - \gamma, \gamma - \alpha\}$;

(iii) $\{\alpha, \alpha + \beta, \alpha + \beta + \gamma\}$.

4. In the space of real $n$-tuples prove that the following vectors are linearly independent:

$$\alpha_1 = (1, 1, 1, \ldots, 1, 1),$$
$$\alpha_2 = (0, 1, 1, \ldots, 1, 1),$$
$$\alpha_3 = (0, 0, 1, \ldots, 1, 1),$$

.

.

.

$$\alpha_n = (0, 0, 0, \ldots, 0, 1).$$

5. In the space $\mathscr{P}_k$ of real polynomials of degree $\leq k$, let $S_i = \{1, x, \ldots, x^i\}$ for $0 \leq i \leq k$. (Here 1 denotes the constant polynomial, $p(x) = 1$.) Use your knowledge about the number of zeros of a polynomial to prove that each $S_i$ is linearly independent. Show also that $[S_k] = \mathscr{P}_k$.

6. The following exercise illustrates that the scalar field of a vector space plays an important role in the concept of linear independence, even though this is not emphasized in the text.

   (i)  Regarding the field of real numbers as a vector space over the field of *rational* numbers, show that $\{1, \sqrt{2}\}$ is linearly *independent*. (The number $\sqrt{2}$ is not rational.)

   (ii)  Regarding the field of real numbers as a vector space over the field of *real* numbers, show that $\{1, \sqrt{2}\}$ is linearly *dependent*.

7. Use Theorem 2.8 to prove Theorem 2.7.

8. In the real space of differentiable functions on $[0, 1]$, determine all functions $f$ such that $\{f, f'\}$ is linearly dependent, where $f'$ is the derivative of $f$.

9. Given that $\mathscr{V} = \mathscr{S} \oplus \mathscr{T}$ (direct sum), prove that if $S$ is a linearly independent subset of $\mathscr{S}$ and $T$ is a linearly independent subset of $\mathscr{T}$, then $S \cup T$ is linearly independent.

10. Prove that if $m > n$, every set of $m$ vectors of $R_n$ is linearly dependent.

## 2.3   Bases and Dimension

When we studied $R_3$ we learned that a choice of basis is equivalent to a choice of coordinate system, and each of our examples of bases in that space contained three vectors. We now consider how these results generalize to arbitrary vector spaces; to begin with, our former definition of a basis can be accepted without change.

**DEFINITION 2.6**   A *basis* for a vector space $\mathcal{V}$ is a linearly independent set that spans $\mathcal{V}$. If $\mathcal{V}$ contains a finite basis, $\mathcal{V}$ is said to be *finite-dimensional*. Otherwise $\mathcal{V}$ is said to be *infinite-dimensional*.

From Theorem 2.7 it follows immediately that every finite-dimensional space has a basis. So does every infinite-dimensional space, but a proof of that fact requires an argument that is more sophisticated logically. There are some important differences between finite-dimensional and infinite-dimensional spaces; we shall observe these distinctions as they arise, but our attention will be focused on finite-dimensional spaces.

**THEOREM 2.9**   If $\mathcal{V} \neq [\theta]$ and if $B$ is any basis for $\mathcal{V}$, every vector of $\mathcal{V}$ has a *unique* representation as a linear combination of the vectors of $B$.

**PROOF**   Suppose that $A = \{\alpha_1, \ldots, \alpha_k\}$ is a basis for $\mathcal{V}$. Since $A$ spans $\mathcal{V}$, each vector $\xi \in \mathcal{V}$ is a linear combination of the basis vectors,

$$\xi = a_1\alpha_1 + \ldots + a_k\alpha_k \qquad \text{for suitable scalars } a_i.$$

Furthermore this representation is *unique*, because if also

$$\xi = b_1\alpha_1 + \ldots + b_k\alpha_k,$$

then

$$\theta = (a_1 - b_1)\alpha_1 + \ldots + (a_k - b_k)\alpha_k.$$

Since $A$ is linearly independent, each coefficient is zero:

$$a_i = b_i \qquad \text{for each } i,$$

which proves the finite-dimensional case of this theorem. The infinite-dimensional case can be proved in essentially the same way.

### Examples of Bases

(a)   A basis for the real vector space $R_n$ of real $n$-tuples is the set $\{\epsilon_1, \epsilon_2, \ldots, \epsilon_n\}$, where $\epsilon_i = (0, \ldots, 1, \ldots, 0)$ with each component 0 except component $i$, which is 1. We later prove that any set of $n$

linearly independent $n$-tuples is a basis; likewise, if a.set of $n$-tuples spans $R_n$ and contains $n$ vectors, it is a basis.

(b)   The space $\mathscr{P}$ of all real polynomials has the infinite set of polynomials $\{1, x, x^2, \ldots, x^n, \ldots\}$ as a basis.

(c)   The field of real numbers regarded as a real vector space has $\{a\}$ as a basis for any $a \neq 0$.

(d)   The field of complex numbers regarded as a real vector space has $\{1, i\}$ as a basis, where $i^2 = -1$.

(e)   The field of complex numbers regarded as a complex vector space has $\{a\}$ as a basis for any $a \neq 0$.

**THEOREM 2.10**   If $\mathscr{V}$ is finite-dimensional, then each basis for $\mathscr{V}$ contains the same number of vectors.

**PROOF**   Let $A = \{\alpha_1, \ldots, \alpha_k\}$ and $B = \{\beta_1, \ldots, \beta_m\}$ be bases for $\mathscr{V}$. Then the ordered set $B_1 = \{\alpha_1, \beta_1, \ldots, \beta_m\}$ is linearly dependent. Apply Theorem 2.8 to extract an ordered basis $B_1{}'$ by successively removing from $B_1$ any vector which is a linear combination of vectors that precede it. The ordered set $B_2$, obtained by writing $\alpha_2$ in front of the ordered $B_1{}'$ is again dependent. Again extract an ordered basis $B_2{}'$ by the process of successively removing any vector that is a linear combination of vectors that precede it. Each time an $\alpha$ is introduced, one or more $\beta$'s are removed. If all of the $\beta$'s are removed before $\alpha_k$ is introduced, we have a basis of the form

$$\{\alpha_p, \ldots, \alpha_2, \alpha_1\}$$

for $p < k$, which contradicts the fact that $A$ is linearly independent. Hence there are at least $k$ vectors in $B$; $m \geq k$. By reversing the roles of $A$ and $B$ in the replacement process we obtain $k \geq m$, so equality must hold.

For an infinite-dimensional space it can be proved that any two bases have the same cardinal number, meaning that there is always a one-to-one correspondence between the elements of two bases for the same space. Theorem 2.10 is important because it permits us to attach a well defined nonnegative integer to each finite-dimensional space. This integer, the number of vectors in any basis of $\mathscr{V}$, is called the *dimension* of $\mathscr{V}$, and in familiar examples it coincides with our geometric idea of dimension.

**DEFINITION 2.7**   The *dimension*, dim $\mathscr{V}$, of a finite-dimensional vector space $\mathscr{V}$ is the number of vectors in any basis of $\mathscr{V}$. Usually, an $n$-dimensional space will be denoted $\mathscr{V}_n$.

Theorems 2.6 and 2.10 together give us a method of obtaining a basis for any finite-dimensional space $\mathscr{V}$. For $\mathscr{V} \neq [\theta]$ begin with any nonzero vector $\alpha_1$; if $\mathscr{V} \neq [\alpha_1]$, choose any $\alpha_2 \in \mathscr{V}$ such that $\alpha_2 \notin [\alpha_1]$. Since $\mathscr{V}$ is finite-dimensional, a finite number of choices by this procedure will produce a basis for $\mathscr{V}$. In short, *any linearly independent set can be extended to a basis*.

Definition 2.6 specifies that a basis for any vector space $\mathscr{V}$ must have two properties: it must be linearly independent, *and* it must span $\mathscr{V}$. But in an $n$-dimensional space a set of $n$ vectors either has both of these properties or neither of them.

**THEOREM 2.11**   Let $A = \{\alpha_1, \ldots, \alpha_n\} \subseteq \mathscr{V}_n$. Then $A$ is linearly independent if and only if $[A] = \mathscr{V}_n$.
PROOF   Exercise. Use Theorems 2.6 and 2.7.

**THEOREM 2.12**   If $\mathscr{S}$ is a subspace of $\mathscr{V}_n$, and if $A$ is a basis for $\mathscr{S}$, then there exists a set $B$ of vectors such that $A \cup B$ is a basis for $\mathscr{V}_n$.
PROOF   Exercise.

**THEOREM 2.13**   If $\mathscr{S}$ and $\mathscr{T}$ are subspaces of $\mathscr{V}_n$, then
$$\dim (\mathscr{S} + \mathscr{T}) + \dim (\mathscr{S} \cap \mathscr{T}) = \dim \mathscr{S} + \dim \mathscr{T}.$$
PROOF   Of the four subspaces involved in this theorem, $\mathscr{S} \cap \mathscr{T}$ is a subspace of each of the others, and both $\mathscr{S}$ and $\mathscr{T}$ are subspaces of $\mathscr{S} + \mathscr{T}$. Let $A = \{\alpha_1, \ldots, \alpha_k\}$ be a basis for $\mathscr{S} \cap \mathscr{T}$. Extend $A$ to a basis $B = \{\alpha_1, \ldots, \alpha_k, \beta_1, \ldots, \beta_i\}$ for $\mathscr{S}$, where $i \geq 0$. Likewise extend $A$ to a basis $C = \{\alpha_1, \ldots, \alpha_k, \gamma_1, \ldots, \gamma_j\}$ for $\mathscr{T}$, where $j \geq 0$. Then $\dim (\mathscr{S} \cap \mathscr{T}) = k$, $\dim \mathscr{S} = k + i$, and $\dim \mathscr{T} = k + j$. The theorem will be proved if we exhibit a basis for $\mathscr{S} + \mathscr{T}$ containing $k + i + j$ vectors. The most obvious choice to try is the set

$$D = \{\alpha_1, \ldots, \alpha_k, \beta_1, \ldots, \beta_i, \gamma_1, \ldots, \gamma_j\},$$

since all are vectors of $\mathscr{S} + \mathscr{T}$. Since we do not know the dimension of $\mathscr{S} + \mathscr{T}$ we must prove that $D$ is linearly independent *and* that $D$ spans $\mathscr{S} + \mathscr{T}$. To prove independence, suppose that

$$a_1\alpha_1 + \ldots + a_k\alpha_k + b_1\beta_1 + \ldots + b_i\beta_i + c_1\gamma_1 + \ldots + c_j\gamma_j = \theta.$$

Let

$$\alpha = \sum_{m=1}^{k} a_m\alpha_m, \qquad \beta = \sum_{m=1}^{i} b_m\beta_m, \qquad \gamma = \sum_{m=1}^{j} c_m\gamma_m.$$

Then

$$\alpha + \beta + \gamma = \theta \qquad \text{where } \alpha \in \mathscr{S} \cap \mathscr{T}, \beta \in \mathscr{S}, \gamma \in \mathscr{T}.$$

Then $\gamma \in \mathscr{S}$, since $\alpha + \beta \in \mathscr{S}$ and $\gamma = -(\alpha + \beta)$. But $\gamma \in \mathscr{T}$, so $\gamma \in \mathscr{S} \cap \mathscr{T}$. Then $\gamma = c_1\gamma_1 + \ldots + c_j\gamma_j$ is a linear combination of the $\alpha$'s, which implies that each $c_r = 0$ since $C$ is linearly independent. Thus $\gamma = \theta$ and $\alpha + \beta = \theta$. But then each $b_s = 0$ and each $a_t = 0$ since $B$ is linearly independent. To show that $D$ spans $\mathscr{S} + \mathscr{T}$, recall that each vector $\xi \in \mathscr{S} + \mathscr{T}$ is of the form

$$\xi = \sigma + \tau \qquad \text{for suitable } \sigma \in \mathscr{S} \text{ and } \tau \in \mathscr{T}.$$

But $\sigma$ is a linear combination of the $\alpha$'s and $\beta$'s, and $\tau$ is a linear combination of $\alpha$'s and $\gamma$'s. Thus $\xi$ is a linear combination of $\alpha$'s, $\beta$'s, and $\gamma$'s.

The relation between bases and coordinate systems is easy to derive from Theorem 2.9. Suppose $A = \{\alpha_1, \ldots, \alpha_k\}$ is a basis for $\mathscr{V}_k$. Each vector $\xi \in \mathscr{V}$ can be represented uniquely as a linear combination of basis vectors.

$$\xi = \sum_{i=1}^{k} a_i\alpha_i.$$

Hence, relative to $A$, $\xi$ determines the unique set of scalars $\{a_1, \ldots, a_k\}$. If we agree to write the basis vectors in a fixed order, for example as indicated by their subscripts, then the set of scalars determined by $\xi$ can be regarded as an ordered $k$-tuple

$$(a_1, a_2, \ldots, a_k).$$

In this way each *ordered* basis for $\mathscr{V}_k$ provides a coordinate system for $\mathscr{V}_k$. Since the ordered basis contains $k$ vectors, each vector of $\mathscr{V}_k$ can be represented uniquely as an ordered $k$-tuple of scalars; conversely any ordered $k$-tuple of scalars $(c_1, \ldots, c_k)$ uniquely determines the vector $c_1\alpha_1 + \ldots + c_k\alpha_k$ of $\mathscr{V}_k$.

In Section 1.5 we agreed to distinguish representations of vectors of $R_3$ relative to the standard basis from representations relative to another basis by writing

$$(a_1, a_2, a_3) \quad \text{for } a_1\epsilon_1 + a_2\epsilon_2 + a_3\epsilon_3,$$

and

$$[a_1, a_2, a_3] \quad \text{for } a_1\alpha_1 + a_2\alpha_2 + a_3\alpha_3.$$

Now that we realize that a $k$-tuple representation of a vector in $\mathcal{V}_k$ depends entirely upon a choice of basis, we do not need to perpetuate this distinction in notation. As before, a $k$-tuple of real numbers $(a_1, \ldots, a_k)$ should be interpreted in terms of the $\epsilon$-basis for $R_k$ unless the context makes clear that a different basis is intended.

Thus, given any real finite-dimensional vector space $\mathcal{V}_k$ and any ordered basis for $\mathcal{V}_k$, a one-to-one correspondence $f$ can be defined from $\mathcal{V}_k$ onto the vector space $R_k$ of all ordered $k$-tuples of elements of $R$ by the rule

$$f\left(\sum_{i=1}^{k} c_i\alpha_i\right) = (c_1, \ldots, c_k).$$

It is easy to verify that $f$ preserves the vector space operations:

$$f(\xi + \eta) = f(\xi) + f(\eta),$$

$$f(a\xi) = af(\xi),$$

for all $\xi$ and $\eta \in \mathcal{V}_k$ and all $a \in R$. This means that the two vector spaces $\mathcal{V}_k$ and $R_k$ are identical in all essential respects and can be distinguished only by the names or notation adopted for each.


**DEFINITION 2.8**   Given two algebraic systems $\mathscr{S}_1$ and $\mathscr{S}_2$, and given a one-to-one correspondence between their operations, a mapping $f$ of the elements of $\mathscr{S}_1$ into the elements of $\mathscr{S}_2$ is called a *homomorphism* provided that

$$f(a * b) = f(a) \circledast f(b)$$

for all $a$ and $b \in \mathscr{S}_1$ and for all pairs of corresponding operations, $*$ in $\mathscr{S}_1$ and $\circledast$ in $\mathscr{S}_2$. In addition, if $f$ is one-to-one, $f$ is called an *isomorphism*, either *into* $\mathscr{S}_2$ or *onto* $\mathscr{S}_2$, depending on the range of $f$.

It should be clear that the previous argument is valid for a finite-dimensional vector space over any field. Hence we can state these results as follows.

**THEOREM 2.14**     Any vector space $\mathscr{V}$ of dimension $k$ over a field $F$ is isomorphic to the vector space $F_k$ of all ordered $k$-tuples of elements of $F$.

From this theorem it is evident that our attempt to gain generality by considering abstract vector spaces was a failure at least for finite-dimensional spaces. We might just as well have considered only spaces of ordered $k$-tuples of field elements, since all vector spaces over the same field $F$ and of the same dimension $k$ are isomorphic to each other and to $F_k$. But $k$-tuples are unnecessarily cumbersome for most of our considerations; even though we have not achieved more generality, we have at least gained convenience in notation. Other advantages will appear as our study progresses.

## EXERCISES 2.3

1. Given the following four vectors of $R_4$ determine the dimension of the subspace that they span, showing your reasoning.

$$\alpha_1 = (0, 1, 1, 2), \quad \alpha_3 = (-2, 1, 0, 1),$$
$$\alpha_2 = (3, 1, 5, 2), \quad \alpha_4 = (1, 0, 3, -1).$$

2. Determine the dimensions of the subspaces given in Exercises 6ii, iv, v of Section 2.1.

3. Let $\mathscr{S}$ be a $k$-dimensional subspace of $\mathscr{V}$, and let $\{\alpha_1, \ldots, \alpha_k\}$ be a basis for $\mathscr{S}$. Prove that each of the following sets is also a basis for $\mathscr{S}$.

  (i) $\{c_1\alpha_1, \ldots, c_k\alpha_k\}$     where $c_1, \ldots, c_k$ are any nonzero scalars.

  (ii) $\{\beta_1, \ldots, \beta_k\}$     where $\beta_i = \alpha_i + \alpha_1$ for $i = 1, \ldots, k$.

4. Let $\mathscr{S}$ be any $k$-dimensional subspace of a vector space $\mathscr{V}$, and let $\{\alpha_1, \ldots, \alpha_k\}$ be a basis for $\mathscr{S}$. Let $\beta \in \mathscr{V}$ but $\beta \notin \mathscr{S}$.

  (i) Prove that the set $\{\alpha_1 - \beta, \alpha_2 - \beta, \ldots, \alpha_k - \beta\}$ is linearly independent.

  (ii) If $\mathscr{T}$ is the $k$-dimensional subspace spanned by the vectors described in (i), determine the dimensions of $\mathscr{S} \cap \mathscr{T}$ and $\mathscr{S} + \mathscr{T}$.

5. Show that if subspaces $\mathscr{T}$ and $\mathscr{S}$ have the same dimension and if $\mathscr{T} \subseteq \mathscr{S}$, then $\mathscr{T} = \mathscr{S}$.

6. Given the basis $\alpha_1 = (1, 1, 1, 1)$, $\alpha_2 = (0, 1, 1, 1)$, $\alpha_3 = (0, 0, 1, 1)$, $\alpha_4 = (0, 0, 0, 1)$, express each vector $\epsilon_i$, $i = 1, 2, 3, 4$, as a linear combination of the $\alpha$'s. Likewise express each $\alpha_i$ as a linear combination of the $\epsilon$'s.

7. Prove Theorem 2.11.

8. Prove Theorem 2.12.

9. Explain why a basis for $\mathscr{V}_n$ can be regarded as a minimal spanning set for $\mathscr{V}_n$, and also as a maximal linearly independent subset of $\mathscr{V}_n$.

10. Let $\mathscr{V}$ be a finite-dimensional vector space.

(i) Prove that if $\{\alpha_1, \ldots, \alpha_n\}$ is a basis for $\mathscr{V}$, if $\mathscr{S} = [\alpha_1, \ldots, \alpha_k]$, and if $\mathscr{T} = [\alpha_{k+1}, \ldots, \alpha_n]$, then $\mathscr{V} = \mathscr{S} \oplus \mathscr{T}$.

(ii) Prove, conversely, that if $\mathscr{S}$ and $\mathscr{T}$ are any subspaces of $\mathscr{V}$ such that $\mathscr{V} = \mathscr{S} \oplus \mathscr{T}$, if $\{\alpha_1, \ldots, \alpha_k\}$ is a basis for $\mathscr{S}$, and if $\{\beta_1, \ldots, \beta_m\}$ is a basis for $\mathscr{T}$, then $\{\alpha_1, \ldots, \alpha_k, \beta_1, \ldots, \beta_m\}$ is a basis for $\mathscr{V}$.

11. Let $\mathscr{S}$ and $\mathscr{T}$ be subspaces of a finite-dimensional space $\mathscr{V}$ such that $\mathscr{S} \cap \mathscr{T} = [\theta]$ and dim $\mathscr{S}$ + dim $\mathscr{T}$ = dim $\mathscr{V}$. Prove that $\mathscr{V} = \mathscr{S} \oplus \mathscr{T}$.

12. (i) If $F_1$ and $F_2$ are fields such that $F_1 \subseteq F_2$, show that $F_2$ may be regarded as a vector space over $F_1$.

(ii) If $Q$, $R$, and $C$ denote the rational, real, and complex fields respectively, determine the dimension of each of the following vector spaces:

$Q$ regarded as a vector space over $Q$.

$C$ regarded as a vector space over $R$.

$C$ regarded as a vector space over $C$.

$R$ regarded as a vector space over $Q$.

13. Define a function $f$ which is an isomorphism from the polynomial space $\mathscr{P}_k$ to the space $R_{k+1}$. Also define an isomorphism $g$ from the space $C$ of complex numbers over $R$ to the space $R_2$.

14. Let $\{\alpha_1, \ldots, \alpha_n\}$ and $\{\beta_1, \ldots, \beta_n\}$ be two bases for $R_n$. Define the mapping $\mathbf{T}$ of $R_n$ into $R_n$ by the statement,

$$\text{if } \xi = \sum_{i=1}^{n} c_i \alpha_i, \quad \text{then } \mathbf{T}(\xi) = \sum_{i=1}^{n} c_i \beta_i.$$

Verify that

(i) $\mathbf{T}$ maps $\alpha_i$ onto $\beta_i$, $i = 1, \ldots, n$,

(ii) $\mathbf{T}$ is a one-to-one mapping of $R_n$ onto $R_n$,

(iii) $\mathbf{T}(\xi + \eta) = \mathbf{T}(\xi) + \mathbf{T}(\eta)$,  for all $\xi$ and $\eta \in R_n$,

(iv) $\mathbf{T}(k\xi) = k\mathbf{T}(\xi)$.

15. Let $h$ be a vector space homomorphism from $\mathscr{V}_n$ to $\mathscr{W}_n$. Prove that $h$ is an isomorphism if and only if $h(\xi) \neq \theta$ in $\mathscr{W}_n$ for every $\xi \neq \theta$ in $\mathscr{V}_n$.

16. Let $f$ be a vector space isomorphism from $\mathscr{V}$ to $\mathscr{W}$. Show that if $S$ is a linearly independent subset of $\mathscr{V}$, then $f(S)$ is linearly independent in $\mathscr{W}$, where $f(S) = \{f(\sigma) \mid \sigma \in S\}$.

## 2.4   Real Inner Products

By the time one studies linear algebra he has developed basic geometric insight from his past experience with Euclidean geometry of the line and the plane. He has also learned how to use coordinate systems to express geometric concepts by algebraic formulas. An extension of geometric intuition to three-dimensional objects seems to come more readily to some people than to others, but fortunately many algebraic formulas for two-space generalize to three-space in a very natural way. The algebraic extension also suggests how geometric concepts can be defined in higher dimensional spaces, where direct geometric insight is likely to fail. Especially for infinite-dimensional spaces we must depend upon algebraic forms to guide our geometric language and thought.

For example, consider the problem of defining the length $\|\alpha\|$ of a vector $\alpha$ in a real vector space $\mathscr{V}$ in such a way that the space acquires metric properties which reasonably can be termed Euclidean. By Theorem 2.14, if $\mathscr{V}$ is $n$-dimensional, it is isomorphic to $R_n$. Relative to a chosen basis, $\alpha$ can be represented as an $n$-tuple $(a_1, \ldots, a_n)$, and we might define

$$\|\alpha\| = \sqrt{a_1^2 + \ldots + a_n^2},$$

since we know that this formula agrees with Euclidean length using the standard basis for $n = 1, 2, 3$. But if we chose a different basis, $\alpha$ would be represented by a different $n$-tuple $(c_1, \ldots, c_n)$, and we would have to decide whether $\alpha$ has the same length in both coordinate systems; that is,

$$\sqrt{a_1^2 + \ldots + a_n^2} = \sqrt{c_1^2 + \ldots + c_n^2}.$$

For this reason a coordinate approach to metric concepts is awkward even for finite-dimensional space.

Hence practicality demands that we take a more general (abstract) approach to metric properties in arbitrary vector spaces, but we can still be guided by our experience in $E_3$. We observed in Section 1.6 that various metric concepts can be expressed in terms of the single notion of scalar product, and in Exercise 3 of that section we derived three general properties of the scalar product that were expressed intrinsically rather than in terms of coordinates. We take these properties as a reasonable point of departure for the study of metric concepts in any real vector space.

**DEFINITION 2.9**  Let $\mathscr{V}$ be a vector space over $R$. A function that assigns to each pair of vectors $(\alpha, \beta) \in \mathscr{V} \times \mathscr{V}$ a real number $\langle \alpha, \beta \rangle$ is called a *real inner product* on $\mathscr{V}$ if and only if for all $\alpha, \beta, \gamma \in \mathscr{V}$ and all $a, b \in R$,

(a)  $\langle a\alpha + b\beta, \gamma \rangle = a\langle \alpha, \gamma \rangle + b\langle \beta, \gamma \rangle$

(b)  $\langle \alpha, \beta \rangle = \langle \beta, \alpha \rangle$,

(c)  $\langle \alpha, \alpha \rangle$ is positive if $\alpha \neq \theta$.

A real vector space for which a real inner product is defined is called a *Euclidean space*.

We use $\langle \alpha, \beta \rangle$ to denote an arbitrary inner product in place of $\alpha \cdot \beta$, used for the dot product in $E_3$. Condition (a) specifies that an inner product is a linear function of the first component. The symmetry condition (b) implies that a real inner product is also a linear function of the second component. Hence a real inner product is bilinear, symmetric, and positive-definite. Using (a) we deduce that

$$\langle \theta, \gamma \rangle = \langle 0\alpha, \gamma \rangle = 0\langle \alpha, \gamma \rangle = 0 \qquad \text{for all } \gamma.$$

Condition (c) guarantees that the inner product attaches a *positive* real number $\langle \alpha, \alpha \rangle$ to each nonzero vector, and this fact is vital later when we define the length of a vector.

Although we shall not consider complex inner products systematically in this text, it is worth noting that a modification of Definition 2.9 is required for vector spaces over the field of complex numbers. Suppose, for example, that $\langle \alpha, \alpha \rangle$ is positive; then $\langle i\alpha, i\alpha \rangle = -\langle \alpha, \alpha \rangle$, where $i^2 = -1$. Hence Condition (c) is inconsistent with (a) and (b) when complex scalars are used. The remedy in that case is to replace the symmetry condition (b) by a *conjugate-symmetry* condition (b'): $\langle \beta, \alpha \rangle = \overline{\langle \alpha, \beta \rangle}$, where the bar denotes complex conjugate. Together with (a), Condition (b′) implies that a complex inner product is *conjugate-bilinear*, meaning that it is linear in the first component but conjugate-linear in the second:

$$\langle \alpha, b\beta + c\gamma \rangle = \bar{b}\langle \alpha, \beta \rangle + \bar{c}\langle \alpha, \gamma \rangle.$$

Since $\bar{x} = x$ for every real number $x$, conjugate symmetry and conjugate bilinearity reduce to symmetry and bilinearity when only real numbers are used for scalars. From these remarks we might expect the theory of complex inner products to be reasonably parallel to the theory which we shall develop for real inner products, and this is indeed the case. A complex inner product space is called a *unitary space*.

### Examples of Euclidean Spaces

(a) The space $R_n$ of all real $n$-tuples is Euclidean if the inner product is defined to be the standard dot product,

$$\langle (x_1, \ldots, x_n), (y_1, \ldots, y_n) \rangle = x_1 y_1 + \ldots + x_n y_n.$$

(b) The infinite-dimensional space $\mathscr{C}[a, b]$ of all real functions continuous on the interval $a \le t \le b$ is Euclidean if the inner product is defined by

$$\langle f, g \rangle = \int_a^b f(t)\, g(t)\, dt.$$

(c) The space $R_2$ is Euclidean if the inner product is defined by

$$\langle (x_1, x_2), (y_1, y_2) \rangle = x_1 y_1 - 2x_1 y_2 - 2x_2 y_1 + 8x_2 y_2.$$

This example will serve to remind us that not every inner product in $R_n$ automatically assumes the form of the standard dot product. The algebraic form of an inner product depends upon the choice of basis (coordinate system), as we shall see later.

**THEOREM 2.15**   (*The Schwarz Inequality*). In any Euclidean space $\mathscr{V}$

$$\langle \alpha, \beta \rangle^2 \le \langle \alpha, \alpha \rangle \langle \beta, \beta \rangle \qquad \text{for all } \alpha, \beta \in \mathscr{V}.$$

**PROOF**   If either $\alpha = \theta$ or $\beta = \theta$, the conclusion $0 \le 0$ is immediate. Otherwise let $t$ be any real number; for any nonzero vectors $\alpha, \beta$,

$$0 \le \langle \alpha + t\beta, \alpha + t\beta \rangle = \langle \alpha, \alpha + t\beta \rangle + t\langle \beta, \alpha + t\beta \rangle$$

$$= \langle \alpha, \alpha \rangle + t\langle \alpha, \beta \rangle + t\langle \beta, \alpha \rangle + t^2\langle \beta, \beta \rangle$$

$$= \langle \alpha, \alpha \rangle + 2t\langle \alpha, \beta \rangle + t^2\langle \beta, \beta \rangle.$$

This last expression is a real quadratic function of $t$ whose values are never negative; its graph is a parabola which opens upward since $\langle \beta, \beta \rangle > 0$. Hence it has either no real zeros or two coincident zeros – that is, the graph either never intersects the horizontal axis or is tangent to that axis. From the quadratic formula, the corresponding algebraic condition is that the discriminant is not positive:

$$(2 \langle \alpha, \beta \rangle)^2 - 4\langle \alpha, \alpha \rangle \langle \beta, \beta \rangle \leq 0,$$

which yields the assertion of the theorem.

**DEFINITION 2.10**   In a Euclidean space $\mathscr{V}$ the *length* $\|\alpha\|$ of a vector $\alpha$ is defined by

$$\|\alpha\| = \langle \alpha, \alpha \rangle^{1/2}.$$

The *distance* $d(\alpha, \beta)$ between two vectors $\alpha$ and $\beta$ is defined by

$$d(\alpha, \beta) = \|\beta - \alpha\|.$$

**THEOREM 2.16**   In a Euclidean space $\mathscr{V}$ length has the following properties; for all $\alpha$ and $\beta \in \mathscr{V}$

(a)   $\|c\alpha\| = |c| \, \|\alpha\|$      for all real $c$,

(b)   $\|\alpha\| > 0$ if $\alpha \neq \theta$, and $\|\theta\| = 0$,

(c)   $\|\alpha + \beta\| \leq \|\alpha\| + \|\beta\|$.

**PROOF**   We prove only (c), which is called *the triangle inequality* (to see why, draw a figure in $E_2$). First we observe that the Schwarz inequality can be expressed as $| \langle \alpha, b \rangle | \leq \|\alpha\| \, \|\beta\|$. Then

$$\|\alpha + \beta\|^2 = \langle \alpha + \beta, \alpha + \beta \rangle = \langle \alpha, \alpha \rangle + \langle \alpha, \beta \rangle + \langle \beta, \alpha \rangle + \langle \beta, \beta \rangle$$

$$= \|\alpha\|^2 + 2\langle \alpha, \beta \rangle + \|\beta\|^2$$

$$\leq \|\alpha\|^2 + 2| \langle \alpha, \beta \rangle | + \|\beta\|^2$$

$$\leq \|\alpha\|^2 + 2\|\alpha\| \, \|\beta\| + \|\beta\|^2 = (\|\alpha\| + \|\beta\|)^2.$$

**THEOREM 2.17**   In a Euclidean space $\mathscr{V}$ distance has the following three properties; for all $\alpha, \beta, \gamma \in \mathscr{V}$

(a)   $d(\alpha, \beta) = d(\beta, \alpha)$,

(b)   $d(\alpha, \beta) > 0$ if $\alpha \neq \beta$, and $d(\alpha, \alpha) = 0$,

(c)  $d(\alpha, \beta) \leq d(\alpha, \gamma) + d(\gamma, \beta)$.

PROOF  Exercise.

Thus distance between points in any Euclidean space is symmetric; it is positive when the points are distinct, and it satisfies the triangle inequality. Any space in which a concept of distance satisfies these three familiar properties of distance is called a *metric space*. Every Euclidean space, therefore, is a metric space; but not every metric space is Euclidean.

## EXERCISES 2.4

1. In $R_2$ let $\alpha = (x_1, x_2)$ and $\beta = (y_1, y_2)$, and let $f_k$ be defined from $R_2 \times R_2$ to $R$ by

$$f_k(\alpha, \beta) = x_1y_1 + x_1y_2 + x_2y_1 + kx_2y_2, \ k \in R.$$

(i)  Show that $f_k$ is symmetric and bilinear for every $k$.

(ii)  Determine all values of $k$ for which $f_k$ is an inner product.

2. Verify that Examples (b) and (c) are Euclidean spaces.

3. Use the inner products of Examples (a) and (b) together with the Schwarz inequality to deduce that

(i)  $$\left( \sum_{i=1}^{n} x_iy_i \right)^2 \leq \left( \sum_{i=1}^{n} x_i^2 \right) \left( \sum_{i=1}^{n} y_i^2 \right)$$

for all real numbers $x_i$ and $y_i$, $i = 1, \ldots, n$;

(ii)  $$\left( \int_a^b f(x)g(x)dx \right)^2 \leq \int_a^b f^2(x)dx \int_a^b g^2(x)dx$$

for all real functions continuous on $[a, b]$.

4. Prove that in any Euclidean space $\mathscr{V}$, if $\alpha$ and $\beta$ are vectors such that $\langle \alpha, \xi \rangle = \langle \beta, \xi \rangle$ for all $\xi \in \mathscr{V}$, then $\alpha = \beta$.

5. Carry out an alternate proof of the Schwarz inequality by expanding $\langle a\alpha + b\beta, a\alpha + b\beta \rangle$ and then choosing $a = -\langle \alpha, \beta \rangle$ and $b = \langle \alpha, \alpha \rangle$.

6. Prove that equality holds in the Schwarz inequality if and only if the set $\{\alpha, \beta\}$ is linearly dependent.

7. Prove parts (a) and (b) of Theorem 2.16.

8. Show that if $\alpha \neq \theta$, then $\|\alpha\|^{-1}\alpha$ is a vector of unit length.

9. Prove Theorem 2.17.

10. Prove that in any Euclidean space
$$\langle \xi + \eta, \xi + \eta \rangle + \langle \xi - \eta, \xi - \eta \rangle = 2\langle \xi, \xi \rangle + 2\langle \eta, \eta \rangle.$$
What familiar geometric theorem does this equation express?

11. Use the notation of arbitrary Euclidean space to prove the theorem that the midpoints of the sides of any quadrilateral are the vertices of a plane parallelogram.

12. Another approach to metric concepts in a real vector space $\mathscr{V}$ is to define a *norm*, which is a function $N$ from $\mathscr{V}$ to $R$ such that for all $\alpha, \beta \in \mathscr{V}$ and all $a \in R$

(a) $N(a\alpha) = |a| N(\alpha)$,

(b) $N(\alpha) > 0$ if $\alpha \neq \theta$,

(c) $N(\alpha + \beta) \leq N(\alpha) + N(\beta)$.

Thus $N(\alpha) = \langle \alpha, \alpha \rangle^{1/2}$ is a norm in any real inner product space. Verify that the following definitions provide norms for $R_n$:

(i)  $N_1((a_1, \ldots, a_n)) = |a_1| + |a_2| + \ldots + |a_n|$,

(ii)  $N((a_i, \ldots, a_n)) = \max\limits_{1 \leq i \leq n} |a_i|$.

13. Referring to Exercise 12 for the definition of a norm, show that the function $N$, defined by
$$N(f) = \max\limits_{0 \leq x \leq 1} |f(x)|$$
is a norm on the space of all functions continuous on $[0, 1]$.

## 2.5   Orthogonality

In order to define the concept of angle between two nonzero vectors in any Euclidean space, we write the Schwarz inequality in the form

$$\frac{|\langle \alpha, \beta \rangle|}{\|\alpha\| \, \|\beta\|} \leq 1,$$

which shows that $\alpha$ and $\beta$ determine a unique number between $-1$ and $1$. That number, in turn, is the cosine of one and only one angle in the closed interval from $0$ to $\pi$.

**DEFINITION 2.11**   In any Euclidean space the *angle* between two nonzero vectors $\alpha$ and $\beta$ is the angle $\Psi(\alpha, \beta)$ between $0$ and $\pi$ such that

$$\cos \Psi(\alpha, \beta) = \frac{\langle \alpha, \beta \rangle}{\|\alpha\| \, \|\beta\|}.$$

Vectors $\alpha$ and $\beta$ are said to be *orthogonal* if and only if $\langle \alpha, \beta \rangle = 0$. The zero vector $\theta$ is orthogonal to every vector.

In the case of complex inner product spaces, the Schwarz inequality is valid, although one step of our proof would have to be modified. If $\langle \alpha, \beta \rangle$ is complex, of course, we cannot always identify a real angle between $\alpha$ and $\beta$ as we can in the Euclidean case, but the concept of orthogonality remains unchanged, and that is enough for many important considerations in unitary spaces.

**THEOREM 2.18**    In any Euclidean space $\mathscr{V}$, orthogonality has these properties.

(a)    $\alpha$ is orthogonal to every $\beta \in \mathscr{V}$ if and only if $\alpha = \theta$.

(b)    If $\alpha$ is orthogonal to every vector of a nonvoid set $S$, then $\alpha$ is orthogonal to every vector in the subspace spanned by $S$.

(c)    Any set of mutually orthogonal nonzero vectors is linearly independent.

**PROOF**    Exercise. *Mutually orthogonal* means that each two distinct vectors of the set are orthogonal.

Our next objective is to demonstrate how a basis of mutually orthogonal vectors can be constructed for any finite-dimensional Euclidean space. The main tool in this construction is the Gram-Schmidt orthogonalization process, which we studied for $R_3$ in Section 1.6. Given any vector $\alpha \neq \theta$ and any vector $\xi \notin [\alpha]$, we can decompose $\xi$ into the sum of a vector $\rho \in [\alpha]$ and a vector $\eta$ orthogonal to $\alpha$. Let

$$p = \frac{\langle \xi, \alpha \rangle}{\langle \alpha, \alpha \rangle} \in R,$$

$$\rho = p\,\alpha,$$

$$\eta = \xi - \rho.$$

Clearly $\rho \in [\alpha]$, and

$$\begin{aligned}
\langle \eta, \alpha \rangle = \langle \xi - \rho, \alpha \rangle &= \langle \xi, \alpha \rangle - \langle \rho, \alpha \rangle \\
&= \langle \xi, \alpha \rangle - \langle p\,\alpha, \alpha \rangle \\
&= \langle \xi, \alpha \rangle - p\,\langle \alpha, \alpha \rangle = 0.
\end{aligned}$$

Hence $\eta$ is orthogonal to $\alpha$. The vector $\rho$ is called the *orthogonal projection* of $\xi$ on $\alpha$. Note that if $\alpha$ is a *unit vector* (that is, $\|\alpha\| = 1$), then $|p| = |\langle \xi, \alpha \rangle|$ is the length of the orthogonal projection of $\xi$ on $\alpha$.

**THEOREM 2.19**   In a finite-dimensional Euclidean space any nonvoid set of mutually orthogonal nonzero vectors can be extended to a basis of mutually orthogonal vectors.

**PROOF**   Let $S = \{\alpha_1, \ldots, \alpha_k\}$ be a mutually orthogonal set of vectors of $\mathscr{V}_n$. $S$ is linearly independent by Theorem 2.18 (c). If $[S] \neq \mathscr{V}_n$, let $\xi \in \mathscr{V}_n$, $\xi \notin [S]$. Let $\rho$ be the sum of the orthogonal projections of $\xi$ on each $\alpha_i$; that is, let

$$p_i = \frac{\langle \xi, \alpha_i \rangle}{\langle \alpha_i, \alpha_i \rangle}, \quad i = 1, \ldots, k,$$

$$\rho_i = p_i \alpha_i, \quad i = 1, \ldots, k,$$

$$\rho = \sum_{i=1}^{k} \rho_i,$$

$$\eta = \xi - \rho.$$

Then for each $j = 1, \ldots, k$

$$\langle \eta, \alpha_j \rangle = \langle \xi - \rho, \alpha_j \rangle = \langle \xi, \alpha_j \rangle - \left\langle \sum_{i=1}^{k} \rho_i, \alpha_j \right\rangle$$

$$= \langle \xi, \alpha_j \rangle - \sum_{i=1}^{k} p_i \langle \alpha_i, \alpha_j \rangle$$

$$= \langle \xi, \alpha_j \rangle - p_j \langle \alpha_j, \alpha_j \rangle,$$

since $\langle \alpha_i, \alpha_j \rangle = 0$ if $i \neq j$. Then the definition of $p_j$ shows that $\langle \eta, \alpha_j \rangle = 0$. Hence $\eta$ is orthogonal to each $\alpha_j$ and hence is orthogonal to every vector in $[S]$. Then $S \cup \{\eta\}$ is a mutually orthogonal extension of $S$. The process can be repeated until a basis for $\mathscr{V}_n$ is obtained.

**DEFINITION 2.12**   In a Euclidean space a set of mutually orthogonal vectors is called an *orthogonal* set. If each vector of an orthogonal set is of unit length, the set is called a *normal orthogonal* (or *orthonormal*) set.

The vector $\rho$, as constructed above, is called the *orthogonal projection* of $\xi$ on the subspace $[S]$. If each $\alpha_i$ is of unit length, then $p_i = \langle \xi, \alpha_i \rangle = \|\xi\| \cos \Psi (\xi, \alpha_i)$, so the numbers $p_1, \ldots, p_k$ are called *direction num-*

*bers* of $\xi$ relative to $[S]$. We also note that the set $\{\alpha_1, \ldots, \alpha_k\}$ is orthonormal if and only if $\langle \alpha_i, \alpha_j \rangle = \delta_{ij}$, where the scalar $\delta_{ij}$, called the *Kronecker delta*, is defined for $i, j = 1, \ldots, k$ by

$$\delta_{ij} = \begin{cases} 1 \text{ if } i = j \\ 0 \text{ if } i \neq j \end{cases}.$$

**THEOREM 2.20**   Every finite-dimensional Euclidean space $\mathscr{V}_n$ has a normal orthogonal basis.

PROOF   We can begin with any $\alpha_1 \neq \theta$ and use the Gram-Schmidt process successively as in Theorem 2.19 to construct an orthogonal set $\{\alpha_1, \ldots, \alpha_n\}$ that must be a basis. Then, letting $\beta_i = \|\alpha_i\|^{-1}\alpha_i$ for $i = 1, \ldots, n$, we obtain $\{\beta_1, \ldots, \beta_n\}$ as a normal orthogonal basis for $\mathscr{V}_n$.

**THEOREM 2.21**   Let $\{\alpha_1, \ldots, \alpha_n\}$ be a normal orthogonal basis for a Euclidean space $\mathscr{V}_n$. If $\xi = \sum\limits_{i=1}^{n} x_i\alpha_i$ and $\eta = \sum\limits_{j=1}^{n} y_j\alpha_j$, then

$$\langle \xi, \eta \rangle = \sum_{i=1}^{n} x_i y_i.$$

PROOF   $\langle \xi, \eta \rangle = \left\langle \sum\limits_{i=1}^{n} x_i\alpha_i, \sum\limits_{j=1}^{n} y_j\alpha_j \right\rangle = \sum\limits_{i=1}^{n} x_i \left( \sum\limits_{j=1}^{n} y_j \langle \alpha_i, \alpha_j \rangle \right)$

$$= \sum_{i=1}^{n} x_i \left( \sum_{j=1}^{n} y_j \delta_{ij} \right) = \sum_{i=1}^{n} x_i y_i.$$

This theorem brings us back to where we started—the standard scalar product in $R_n$. *Any* inner product on a finite-dimensional space must assume this form *relative to a normal orthogonal basis*. This means that in any $n$-dimensional Euclidean space if we use a normal orthogonal basis, then the scalar product, length, distance, and angle assume the familiar forms:

$$\langle \xi, \eta \rangle = x_1 y_1 + \ldots + x_n y_n,$$

$$\|\xi\| = \sqrt{x_1^2 + \ldots + x_n^2},$$

$$d(\xi, \eta) = \sqrt{(x_1 - y_1)^2 + \ldots + (x_n - y_n)^2},$$

$$\cos \Psi(\xi, \eta) = \frac{x_1 y_1 + \ldots + x_n y_n}{\sqrt{x_1^2 + \ldots + x_n^2} \sqrt{y_1^2 + \ldots + y_n^2}}.$$

## EXERCISES 2.5

1. Show that the last formula of this section is equivalent to the trigonometric law of cosines.

2. Prove Theorem 2.18.

3. Prove that in any Euclidean space
$$\langle \xi + \eta, \xi - \eta \rangle = 0 \qquad \text{if and only if } \|\xi\| = \|\eta\|.$$
Restate this result in the language of plane geometry.

4. State and prove the Pythagorean theorem and its converse in the language and notation of arbitrary Euclidean space.

5. Prove that in any Euclidean space $\langle \alpha, \beta \rangle = 0$ if and only if
$$\|\alpha + t\beta\| \geq \|\alpha\| \qquad \text{for every real number } t.$$
Interpret geometrically.

6. In $R_4$ with the standard basis and inner product, let $\alpha_1 = (2, 1, -5, 0)$ and $\alpha_2 = (3, -1, 1, 0)$. Use the Gram-Schmidt process to construct an orthogonal basis $\{\alpha_1, \alpha_2, \alpha_3, \alpha_4\}$. Convert this to a normal orthogonal basis.

7. Let $\{\alpha_1, \ldots, \alpha_n\}$ be a normal orthogonal basis for $E_n$. Prove

  (i)  Bessel's inequality: $\sum_{i=1}^{k} \langle \xi, \alpha_i \rangle^2 \leq \|\xi\|^2$ for all $k \leq n$,

  (ii) Parseval's identity: $\langle \xi, \eta \rangle = \sum_{i=1}^{n} \langle \xi, \alpha_i \rangle \langle \alpha_i, \eta \rangle$.

8. Let $\mathscr{S}$ be a subspace of Euclidean $n$-space. Prove that

  (i)  any linear combination of vectors which are orthogonal to $\mathscr{S}$ is itself orthogonal to $\mathscr{S}$,

  (ii) the only vector of $\mathscr{S}$ which is orthogonal to $\mathscr{S}$ is $\theta$,

  (iii) each vector $\xi$ has a unique decomposition
$$\xi = \sigma + \tau,$$
where $\sigma \in \mathscr{S}$ and $\tau$ is orthogonal to $\mathscr{S}$.

9. Let $\mathscr{S}, \mathscr{T}$ be subspaces of an arbitrary Euclidean space $E$, and let $\mathscr{S}^{\perp} = \{\xi \in E \mid \langle \xi, \sigma \rangle = 0 \text{ for all } \sigma \in \mathscr{S}\}$. $\mathscr{S}^{\perp}$ is called the *orthogonal complement* of $\mathscr{S}$. Prove the following statements.

  (i)   $\mathscr{S}^{\perp}$ is a subspace of $E$.

  (ii)  If $E$ is finite-dimensional, $E = \mathscr{S} \oplus \mathscr{S}^{\perp}$.

  (iii) $\mathscr{S} \subseteq (\mathscr{S}^{\perp})^{\perp}$, and equality holds if $E$ is finite-dimensional.

  (iv)  $(\mathscr{S} + \mathscr{T})^{\perp} = \mathscr{S}^{\perp} \cap \mathscr{T}^{\perp}$.

  (v)   $(\mathscr{S} \cap \mathscr{T})^{\perp} \supseteq \mathscr{S}^{\perp} + \mathscr{T}^{\perp}$, and equality holds if $E$ is finite-dimensional.

10. Consider the polynomial space $\mathscr{P}_2$ with the inner product defined by

$$\langle p, q \rangle = \int_0^1 p(x)q(x)dx.$$

(i) Use the Gram-Schmidt process to convert the standard basis $B = \{1, x, x^2\}$ to the normal orthogonal basis

$$N = \{1, \sqrt{3}(2x - 1), \sqrt{5}(6x^2 - 6x + 1)\}.$$

(ii) Illustrate Theorem 2.21 by expressing each of the two polynomials

$$p(x) = x^2$$
$$q(x) = 2x - 6x^2$$

as a linear combination of the vectors of $N$ and comparing $\langle p, q \rangle$ with the dot product of those two linear combinations.

# 3

# *Linear Mappings and Matrices*

## 3.1   Linear Mappings

We turn now from a study of vector spaces to the central concern of linear algebra—a study of *linear mappings* from one vector space $\mathscr{V}$ over a field $F$ to another vector space $\mathscr{W}$ over the *same* field. The term *mapping* is used instead of function in order to emphasize the geometric nature of this study. The adjective *linear* when applied to a mapping $h$ means that $h$ preserves all linear combinations: for all $\alpha, \beta \in \mathscr{V}$ and all $a, b \in F$

   (1)   $h(a\alpha + b\beta) = ah(\alpha) + bh(\beta)$.

If $a = 1 = b$ we obtain

   (2)   $h(\alpha + \beta) = h(\alpha) + h(\beta)$     for all $\alpha, \beta \in \mathscr{V}$,

and if $b = 0$ we obtain

   (3)   $h(a\alpha) = ah(\alpha)$     for all $\alpha \in \mathscr{V}, a \in F$.

Conversely if (2) and (3) are valid, then so is (1). Hence in the algebraic terminology of Definition 2.8, a linear mapping is simply a *vector space homomorphism*.

It is useful to extract from this variety of terms those that seem to be well suited for this particular study. A linear function between vector spaces $\mathscr{V}$ and $\mathscr{W}$ over the same field $F$ will be called a *linear mapping* and will be denoted by boldface capital letters such as **T** and **S**. In the special case in which $\mathscr{W} = \mathscr{V}$, **T** will be called a *linear transformation of* $\mathscr{V}$. The term *linear operator* is frequently used for a linear transformation of an infinite-dimensional space or a space of functions. In case $\mathscr{W} = F$, linear transformations will be called *linear functionals* and will be denoted by boldface lower case letters such as **f** and **g**. In customary usage the symbol $\mathbf{T}(\alpha)$ denotes the image of $\alpha$ under the mapping **T**. Frequently we are able to simplify this notation by not using the parentheses unless they are needed for clarity. Thus we shall write $\mathbf{T}\alpha$ instead of $\mathbf{T}(\alpha)$, but we will not write $\mathbf{T}\alpha + \beta$ when we mean $\mathbf{T}(\alpha + \beta)$.

**DEFINITION 3.1**   Let $\mathscr{V}$ and $\mathscr{W}$ be vector spaces over a field $F$. A function **T** from $\mathscr{V}$ to $\mathscr{W}$ is called a *linear mapping* if and only if for all $\alpha, \beta \in \mathscr{V}$ and all $a, b \in F$,

$$\mathbf{T}(a\alpha + b\beta) = a\mathbf{T}\alpha + b\mathbf{T}\beta.$$

A linear mapping from $\mathscr{V}$ into $\mathscr{V}$ is called a *linear transformation* on $\mathscr{V}$.

Since linear mappings from $\mathscr{V}$ to $\mathscr{W}$ are functions, we can combine them algebraically as we do functions. Indeed, the set $\mathscr{L}(\mathscr{V}, \mathscr{W})$ of all linear mappings from $\mathscr{V}$ to $\mathscr{W}$ is an algebraic system in which *equality*, *sum*, and *scalar multiple* of linear mappings are defined by

$\mathbf{T}_1 \ominus \mathbf{T}_2$ if and only if $\mathbf{T}_1\alpha = \mathbf{T}_2\alpha$    for all $\alpha \in \mathscr{V}$,

$(\mathbf{T}_1 \oplus \mathbf{T}_2)\alpha = \mathbf{T}_1\alpha + \mathbf{T}_2\alpha$,

$(c \odot \mathbf{T}_1)\alpha = c(\mathbf{T}_1\alpha)$.

The circled symbols refer to relations and operations in $\mathscr{L}(\mathscr{V}, \mathscr{W})$, while the uncircled symbols refer to those in $\mathscr{W}$. As before, the use of distinctive notation for linear mappings permits us to omit the circles without fear of ambiguity.

We must, however, be certain that the sum and scalar multiple of linear mappings are themselves linear; we have

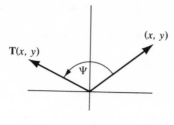

FIGURE 3.1

$$(T_1 + T_2)(a\alpha + b\beta) = T_1(a\alpha + b\beta) + T_2(a\alpha + b\beta)$$
$$= aT_1\alpha + bT_1\beta + aT_2\alpha + bT_2\beta$$
$$= a(T_1 + T_2)\alpha + b(T_1 + T_2)\beta.$$

In this calculation we used the definition of $T_1 + T_2$, the linearity of $T_1$ and $T_2$, associativity and commutativity of addition in $\mathscr{W}$, and the distributivity property of scalar multiples in $\mathscr{W}$. A similar calculation shows that $cT_1$ is linear when $T_1$ is linear.

### Examples of Linear Mappings

(a)  Let $\mathscr{V}$ and $\mathscr{W}$ be any vector spaces over $F$. The *zero* mapping $Z$ is defined by

$$Z\alpha = \theta \qquad \text{for each } \alpha \in \mathscr{V}.$$

$Z$ is indeed linear, and for every $T \in \mathscr{L}(\mathscr{V}, \mathscr{W})$

$$T + Z = T = Z + T.$$

(b)  Let $T \in \mathscr{L}(\mathscr{V}, \mathscr{W})$. The mapping $-T$, defined by

$$(-T)\alpha = -(T\alpha)$$

is also linear, and

$$T + (-T) = Z = (-T) + T.$$

(c)  In $E_2$ the transformation $T_\Psi$ rotates each point of the plane through a counterclockwise angle $\Psi$ about the origin:

$$T_\Psi(x, y) = (x\cos \Psi - y\sin \Psi, \, x\sin \Psi + y\cos \Psi).$$

Direct calculations will verify that $T$ is linear. See Figure 3.1.

(d)   In the space $\mathscr{C}[0, 1]$ let $\mathbf{J}$ be defined for each $f \in \mathscr{C}[0, 1]$ by

$$(\mathbf{J}f)(x) = \int_0^x f(t)dt, \quad \text{for each } x \in [0, 1].$$

Then $\mathbf{J}$ is a linear operator.

(e)   A linear functional $\mathbf{f}$ from $R_3$ to $R$ can be defined by

$$\mathbf{f}((a_1, a_2, a_3)) = a_1 + a_2 + a_3.$$

(f) The definite integral is a linear functional $\mathbf{i}$ on the space of all real functions that are continuous on $[0, 1]$:

$$\mathbf{i}(f) = \int_0^1 f(x)dx.$$

(g)   Let $\alpha$ be a fixed vector in $E_n$, and let $\mathbf{f}_\alpha$ be defined for all $\xi \in E_n$ by

$$\mathbf{f}_\alpha(\xi) = \langle \alpha, \xi \rangle.$$

Then $\mathbf{f}_\alpha$ is a linear functional on $E_n$, and if $\alpha \neq \beta$ then $\mathbf{f}_\alpha \neq \mathbf{f}_\beta$.

**THEOREM 3.1**   Let $\mathscr{V}$ and $\mathscr{W}$ be vector spaces over $F$, and let $L$ denote the set of all linear mappings from $\mathscr{V}$ to $\mathscr{W}$. The system

$$\mathscr{L}(\mathscr{V}, \mathscr{W}) = \{L, F; +, \cdot, \oplus, \odot\}$$

is a vector space over $F$.

PROOF   We have verified that $L$ is closed under the operations $\oplus$ and $\odot$. Examples (a) and (b) show that $\mathbf{Z}$ is the zero element of $\mathscr{L}$ and that $-\mathbf{T}$ is the additive inverse of $\mathbf{T}$. The other postulates of a vector space are satisfied because of the corresponding properties in $\mathscr{W}$ and the definitions of $\oplus$ and $\odot$.

The space $\mathscr{L}(\mathscr{V}, F)$ of all linear functionals (linear mappings from $\mathscr{V}$ to $F$) is called the *dual space* of $\mathscr{V}$. In the finite-dimensional case, the dual space of $\mathscr{V}_n$ is also of dimension $n$ and hence is isomorphic to $\mathscr{V}_n$.

From your study of calculus you will remember that the function of a function concept provides another important way of combining functions. Given functions $f$ and $g$, if the domain of $g$ includes the range of $f$, then the *composition function* $g \circ f$ is defined for each $x$ in the domain of $f$ by the rule

$$(g \circ f)(x) = g(f(x)).$$

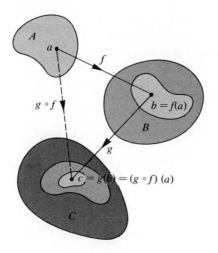

FIGURE 3.2

A mapping diagram is useful here; let $f$ map $A$ into $B$ and let $g$ map $B$ into $C$; then $g \circ f$ maps $A$ into $C$.

In the vector space analogy, we wish to be sure that when the composition of linear mappings is defined, it is also linear. Let **T** be a linear mapping from $\mathscr{V}$ to $\mathscr{W}$ and **S** a linear mapping from $\mathscr{W}$ to $\mathscr{Y}$. Then for all $\alpha, \beta \in \mathscr{V}$ and all $a, b \in F$ the linearity of **T** and of **S** yields

$$(\mathbf{S} \circ \mathbf{T})\,(a\alpha + b\beta) = \mathbf{S}(\mathbf{T}(a\alpha + b\beta)) = \mathbf{S}(a\mathbf{T}\alpha + b\mathbf{T}\beta)$$

$$= a\mathbf{S}(\mathbf{T}\alpha) + b\mathbf{S}(\mathbf{T}\beta) = a(\mathbf{S} \circ \mathbf{T})\alpha + b(\mathbf{S} \circ \mathbf{T})\beta,$$

so $\mathbf{S} \circ \mathbf{T}$ is linear.

**DEFINITION 3.2** Let $\mathscr{V}$, $\mathscr{W}$, and $\mathscr{Y}$ be vector spaces over $F$. Let **T** be a linear mapping from $\mathscr{V}$ to $\mathscr{W}$, and let **S** be a linear mapping from $\mathscr{W}$ to $\mathscr{Y}$. Then the *product* $\mathbf{S} \circ \mathbf{T}$ is the linear mapping from $\mathscr{V}$ to $\mathscr{Y}$ defined for all $\alpha \in \mathscr{V}$ by

$$(\mathbf{S} \circ \mathbf{T})\,(\alpha) = \mathbf{S}(\mathbf{T}\alpha).$$

Normally we shall denote $\mathbf{S} \circ \mathbf{T}$ by $\mathbf{ST}$.

Note that if **S** and **T** are linear *transformations* on $\mathscr{V}$, then the products **ST** and **TS** are both defined, though our experience with composition of functions warns us not to expect **ST** and **TS** to be the same mapping. However, as an exercise you may verify that the product of linear transformations is associative and bilinear:

$$\mathbf{T}_3(\mathbf{T}_2\mathbf{T}_1) = (\mathbf{T}_3\mathbf{T}_2)\mathbf{T}_1,$$

$$\mathbf{T}_3(a\,\mathbf{T}_2 + b\,\mathbf{T}_1) = a\,\mathbf{T}_3\mathbf{T}_2 + b\,\mathbf{T}_3\mathbf{T}_1,$$

$$(a\,\mathbf{T}_3 + b\,\mathbf{T}_2)\mathbf{T}_1 = a\,\mathbf{T}_3\mathbf{T}_1 + b\,\mathbf{T}_2\mathbf{T}_1.$$

Hence the system $\mathscr{L}(\mathscr{V}) = \{L, F; +, \cdot, \oplus, \odot, \circ\}$ of all linear transformations from $\mathscr{V}$ into $\mathscr{V}$ is a vector space for which a binary operation, $\circ$, is defined on the elements of $L$. Furthermore, that operation is closed, associative and bilinear. Such a system is called a *linear algebra*. In a linear algebra the notation $\mathbf{T}^k$ means $\mathbf{T}\mathbf{T} \ldots \mathbf{T}$ ($k$ times).

### Further Examples of Linear Mappings

(h) For any vector space $\mathscr{V}$ the transformation **I** defined by

$$\mathbf{I}\alpha = \alpha \qquad \text{for each } \alpha \in \mathscr{V}$$

is linear. It is called the *identity* mapping and has the property that for any linear transformation **T** on $\mathscr{V}$

$$\mathbf{TI} = \mathbf{T} = \mathbf{IT}.$$

(i) In the polynomial space $\mathscr{P}_n$, the derivative operator **D**, defined by

$$\mathbf{D}p(x) = \frac{d}{dx}\,p(x),$$

is linear. For $n \geq 1$, $\mathbf{D} \neq \mathbf{Z}$, but $\mathbf{D}^{n+1} = \mathbf{Z}$, since the $(n+1)$st derivative of any polynomial of degree $\leq n$ is the zero polynomial.

(j) In $R_3$, the transformations **T** and **S**, defined by

$$\mathbf{T}(a_1, a_2, a_3) = (a_2, a_1, a_3)$$

$$\mathbf{S}(a_1, a_2, a_3) = (0, a_2, a_3)$$

are linear. You may verify that $\mathbf{ST} \neq \mathbf{TS}$, $\mathbf{STS} = \mathbf{STST}$, $\mathbf{S}^2 = \mathbf{S}$, and $\mathbf{T}^2 = \mathbf{I}$. Geometrically **T** reflects each point horizontally across the vertical plane $x = y$, and **S** projects each point orthogonally onto the vertical plane $x = 0$.

From these examples it should be evident that the algebra of linear mappings differs from the familiar algebra of numbers. In particular, the behavior exhibited by Examples (i) and (j) is important enough in our future work to merit special terminology.

**DEFINITION 3.3**   Let **T** be a linear transformation on $\mathscr{V}$.

(a)   **T** is said to be *nilpotent* if and only if there exists an integer $k > 1$ such that $\mathbf{T}^k = \mathbf{Z}$ but $\mathbf{T}^{k-1} \neq \mathbf{Z}$. The integer $k$ is called the *index of nilpotency* of **T**.

(b)   **T** is said to be *idempotent* if and only if $\mathbf{T}^2 = \mathbf{T}$.

Finally we observe that linear mappings from $\mathscr{V}$ to $\mathscr{W}$ are very easy to construct provided that we have a basis $B$ for $\mathscr{V}$. For each $\beta_i \in B$, we can choose *any* vector $\gamma_i \in \mathscr{W}$ and define $\mathbf{T}\beta_i = \gamma_i$. However, if **T** is to be linear, we have no further choice, because for each $\xi \in \mathscr{V}$,

$$\text{if } \xi = \sum_{i=1}^{p} b_i \beta_i, \quad \text{then } \mathbf{T}\xi = \sum_{i=1}^{p} b_i \mathbf{T}\beta_i.$$

Thus a linear mapping is *completely* determined by its effect on the vectors of any basis of its domain space.

**EXERCISES** 3.1

1. Prove that $\mathbf{T}\theta = \theta$ for any linear mapping **T** from $\mathscr{V}$ to $\mathscr{W}$.

2. Prove that $\mathbf{TZ} = \mathbf{Z} = \mathbf{ZT}$ for any linear transformation **T** on $\mathscr{V}$, where **Z** denotes the zero transformation on $\mathscr{V}$.

3. (i)   Is the mapping **J** of Example (d) a linear operator on the polynomial space $\mathscr{P}_2$? Explain.

(ii)   Is the mapping **D** of example (i) a nilpotent linear operator on the space $\mathscr{P}$ of all real polynomials? Explain.

4. Verify that the rotation transformation $\mathbf{T}_\Psi$ is linear on $E_2$. Both algebraically and geometrically show that $\mathbf{T}_\omega \mathbf{T}_\Psi = \mathbf{T}_{\omega + \Psi}$.

5. A linear transformation **T** of the Cartesian plane maps $(1, 2)$ into $(-2, 1)$, and also maps $(-1, 1)$ into $(5, -7)$. Compute the image of $(x, y)$. (*Hint:* $\mathbf{T}(\epsilon_1 + 2\epsilon_2) = -2\epsilon_1 + \epsilon_2$ and $\mathbf{T}(-\epsilon_1 + \epsilon_2) = 5\epsilon_1 - 7\epsilon_2$. Determine $\mathbf{T}(x\epsilon_1 + y\epsilon_2)$.)

6. Let $h$ denote the conjugate function for complex numbers,

$$h(x + iy) = x - iy.$$

(i)  Considering the complex numbers as a vector space over the field of real numbers, is $h$ a linear transformation?

(ii)  Considering the complex numbers as a vector space over the field of complex numbers, is $h$ a linear transformation?

(iii)  In each case state which of the two linearity properties (2) and (3) is satisfied by $h$.

7.  Find an example of a mapping $h$ of $R_2$ that satisfies Property (3) of linearity but does not satisfy Property (2).

8.  Let $\mathbf{T}$, $\mathbf{S}$, and $\mathbf{R}$ be defined from $R_2$ to $R_2$ by

$$\mathbf{T}(x, y) = (y, x),$$
$$\mathbf{S}(x, y) = (x, 0),$$
$$\mathbf{R}(x, y) = (0, y).$$

Describe each geometrically and show that each is linear. Also show that $\mathbf{RS} = \mathbf{SR} = \mathbf{Z}$, $\mathbf{T}^2 = \mathbf{I}$, $\mathbf{TS} \neq \mathbf{ST}$, and $\mathbf{TST} = \mathbf{R}$.

9.  Prove that the product (composition) of linear mappings is associative and bilinear.

10.  In the space $\mathscr{P}$ of all real polynomials of all degrees, let $\mathbf{D}$ and $\mathbf{M}$ be the mappings defined by

$$\mathbf{D}p(x) = \frac{d}{dx} p(x),$$

$$\mathbf{M}p(x) = xp(x).$$

(i)  Show that $\mathbf{D}$ and $\mathbf{M}$ are linear transformations on $\mathscr{P}$.

(ii)  Show that $\mathbf{DM} - \mathbf{MD} = \mathbf{I}$.

(iii)  Show that $(\mathbf{MD})^2 = \mathbf{M}^2\mathbf{D}^2 + \mathbf{MD}$.

11.  Let $\mathbf{T}$ be a linear transformation on $\mathscr{V}$ that is nilpotent of index $p$. Show that if $\mathbf{T}^{p-1}\xi \neq \theta$, then $\{\xi, \mathbf{T}\xi, \ldots, \mathbf{T}^{p-1}\xi\}$ is linearly independent.

12.  Generalize Example (e) as follows. For all vectors $(x_1, \ldots, x_n)$ in $R_n$ define the function $\mathbf{f}$ by

$$\mathbf{f}((x_1, \ldots, x_n)) = c_1x_1 + c_2x_2 + \ldots + c_nx_n,$$

where the numbers $c_1, c_2, \ldots, c_n$ are arbitrary but fixed. Prove that $\mathbf{f}$ is a linear functional on $R_n$. How is this example related to Example (g)?

13.  Let $\mathscr{V}$ be a real vector space on which a norm $N$ is defined. (See Exercise 12 of Section 2.4.) Show that the space $\mathscr{L}(\mathscr{V})$ of all linear transformations on $\mathscr{V}$ can be made into a normed space by defining for each $\mathbf{T} \in \mathscr{L}(\mathscr{V})$

$$N_1(\mathbf{T}) = \max_{N(\xi)=1} N(\mathbf{T}\xi);$$

that is, show that $N_1$ satisfies the three defining properties of a norm.

## 3.2   Rank and Nullity

One of the most effective ways to study a given linear mapping is to observe its effect on various subspaces, particularly those subspaces that **T** itself defines in a natural way. We shall first prove that any linear mapping **T** of a vector space $\mathscr{V}$ into a vector space $\mathscr{W}$ carries each subspace of $\mathscr{V}$ into a subspace of $\mathscr{W}$. In particular $\mathbf{T}(\mathscr{V})$ is a subspace of $\mathscr{W}$. Furthermore, any subspace of $\mathbf{T}(\mathscr{V})$ is the T-image of a subspace of $\mathscr{V}$. Then we focus our attention on two subspaces (one subspace of $\mathscr{V}$ and one of $\mathscr{W}$), which are intrinsically related to **T**.

**THEOREM 3.2**   Let **T** be a linear mapping from $\mathscr{V}$ into $\mathscr{W}$. If $\mathscr{S}$ is any subspace of $\mathscr{V}$, then $\mathbf{T}(\mathscr{S})$ is a subspace of $\mathscr{W}$, where

$$\mathbf{T}(\mathscr{S}) = \{\mathbf{T}\sigma \mid \sigma \in \mathscr{S}\}.$$

Moreover, if $\mathscr{Y}$ is any subspace of $\mathbf{T}(\mathscr{V})$ in $\mathscr{W}$, then the set

$$\{\xi \in \mathscr{V} \mid \mathbf{T}\xi \in \mathscr{Y}\}$$

is a subspace of $\mathscr{V}$.

**PROOF**    Let $\mathscr{S}$ be a subspace of $\mathscr{V}$. $\mathbf{T}(\mathscr{S})$ is nonvoid since $\theta \in \mathbf{T}(\mathscr{S})$, so by Theorem 2.2 $\mathbf{T}(\mathscr{S})$ is a subspace if it is closed under vector sum and scalar multiple. If $\beta, \eta \in \mathbf{T}(\mathscr{S})$ then $\beta = \mathbf{T}\alpha$ for some $\alpha \in \mathscr{S}$, and $\eta = \mathbf{T}\xi$ for some $\xi \in \mathscr{S}$. Hence $\beta + \eta = \mathbf{T}\alpha + \mathbf{T}\xi = \mathbf{T}(\alpha + \xi)$. But $\alpha + \xi \in \mathscr{S}$ since $\mathscr{S}$ is a subspace of $\mathscr{V}$, so $\beta + \eta \in \mathbf{T}(\mathscr{S})$. Furthermore, for $c \in F$, $c\beta = c\mathbf{T}\alpha = \mathbf{T}(c\alpha) \in \mathbf{T}(\mathscr{S})$, since $\mathscr{S}$ is a subspace. Hence the T-image of every subspace of $\mathscr{V}$ is a subspace of $\mathscr{W}$. In particular the range $\mathbf{T}(\mathscr{V})$ of **T** is a subspace of $\mathscr{W}$.

Now, if $\mathscr{Y}$ is any subspace of $\mathbf{T}(\mathscr{V})$, let $X = \{\xi \in \mathscr{V} \mid \mathbf{T}\xi \in \mathscr{Y}\}$. $X$ is nonvoid since $\theta \in X$. If $\gamma, \delta \in X$, then $\mathbf{T}\gamma, \mathbf{T}\delta \in \mathscr{Y}$, and $\mathbf{T}(\gamma + \delta) = \mathbf{T}\gamma + \mathbf{T}\delta \in \mathscr{Y}$ since $\mathscr{Y}$ is a subspace. Also $\mathbf{T}(c\gamma) = c\mathbf{T}\gamma \in \mathscr{Y}$, so $X$ is a subspace of $\mathscr{V}$.

**DEFINITION 3.4**   Let **T** be a linear mapping of $\mathscr{V}$ into $\mathscr{W}$.

(a)   The *range space* (or the *image space*) of **T** is the subspace $\mathscr{R}(\mathbf{T}) = \{\mathbf{T}\xi \in \mathscr{W} \mid \xi \in \mathscr{V}\}$. If $\mathscr{R}(\mathbf{T})$ is finite-dimensional, dim $\mathscr{R}(\mathbf{T})$ is called the *rank* of **T** and is denoted $r(\mathbf{T})$.

(b)   The *null space* of **T** is the subspace $\mathcal{N}(\mathbf{T}) = \{\xi \in \mathcal{V} \mid \mathbf{T}\xi = \theta\}$. If $\mathcal{N}(\mathbf{T})$ is finite-dimensional, dim $\mathcal{N}(\mathbf{T})$ is called the *nullity* of **T** and is denoted $n(\mathbf{T})$. $\mathcal{N}(\mathbf{T})$ is also called the kernel of **T**.

Thus each linear mapping **T** from $\mathcal{V}$ to $\mathcal{W}$ determines the subspace $\mathcal{N}(\mathbf{T}) \subseteq \mathcal{V}$ of all vectors mapped by **T** into $\theta \in \mathcal{W}$ and the subspace $\mathcal{R}(\mathbf{T}) \subseteq \mathcal{W}$ of all images of the mapping. These subspaces and their dimensions are important concepts in the study of linear mappings.

For example, the second derivative operator $\mathbf{D}^2$ maps the polynomial space $\mathcal{P}_n$ into itself and onto the subspace $\mathcal{P}_{n-2}$. Thus the range space of $\mathbf{D}^2$ is $\mathcal{P}_{n-2}$. The null space of $\mathbf{D}^2$ consists of all polynomials that are carried into the zero polynomial by $\mathbf{D}^2$. Hence the null space of $\mathbf{D}^2$ is $\mathcal{P}_1$. Furthermore,

$$r(\mathbf{D}^2) + n(\mathbf{D}^2) = (n-1) + (2) = \dim(\mathcal{P}_n).$$

In this case at least the sum of the rank and the nullity of $\mathbf{D}^2$ is the dimension of $\mathcal{P}_n$. The next two theorems prove that this observation is valid for any linear transformation on any $\mathcal{V}_n$.

**THEOREM 3.3**   Let **T** be a linear mapping from an $n$-dimensional vector space $\mathcal{V}$ to a vector space $\mathcal{W}$. If a basis $\{\alpha_1, \ldots, \alpha_k\}$ for $\mathcal{N}(\mathbf{T})$ is extended in any way to a basis $\{\alpha_1, \ldots, \alpha_k, \alpha_{k+1}, \ldots, \alpha_n\}$ for $\mathcal{V}$, then $\{\mathbf{T}\alpha_{k+1}, \ldots, \mathbf{T}\alpha_n\}$ is a basis for $\mathcal{R}(\mathbf{T})$.

**PROOF**   Any vector $\eta$ of $\mathcal{R}(\mathbf{T})$ is of the form $\eta = \mathbf{T}\xi$ for some $\xi \in \mathcal{V}$. Let a basis $\{\alpha_1, \ldots, \alpha_k\}$ for $\mathcal{N}(\mathbf{T})$ extend to a basis $\{\alpha_1, \ldots, \alpha_n\}$ for $\mathcal{V}$, and let $\xi = \sum_{i=1}^{n} a_i\alpha_i$. Then

$$\mathbf{T}\xi = \mathbf{T}\left(\sum_{i=1}^{n} a_i\alpha_i\right) = \sum_{i=1}^{n} a_i\mathbf{T}\alpha_i = \sum_{i=k+1}^{n} a_i\mathbf{T}\alpha_i$$

since $\mathbf{T}\alpha_i = \theta$ for $i \leq k$. Hence $\mathcal{R}(\mathbf{T}) = [\mathbf{T}\alpha_{k+1}, \ldots, \mathbf{T}\alpha_n]$. We must also show that these vectors are linearly independent. Suppose that

$$b_{k+1}\mathbf{T}\alpha_{k+1} + \ldots + b_n\mathbf{T}\alpha_n = \theta.$$

Then $\mathbf{T}(b_{k+1}\alpha_{k+1} + \ldots + b_n\alpha_n) = \theta$ since **T** is linear. Thus

$$\sum_{i=k+1}^{n} b_i\alpha_i \in \mathcal{N}(\mathbf{T}) = [\alpha_1, \ldots, \alpha_k].$$

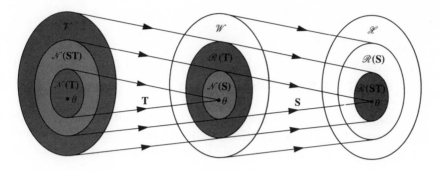

FIGURE 3.3

Hence for suitable scalars $c_j$,

$$\sum_{i=k+1}^{n} b_i \alpha_i = \sum_{j=1}^{k} c_j \alpha_j.$$

Since $\{\alpha_1, \ldots, \alpha_n\}$ is a basis for $\mathscr{V}$, each $b_i$ and each $c_j$ must be zero. Hence $\{T\alpha_{k+1}, \ldots, T\alpha_n\}$ is linearly independent and spans $\mathscr{R}(T)$, as the theorem asserts.

**THEOREM 3.4** If $T$ is a linear mapping of a finite-dimensional space $\mathscr{V}$ into $\mathscr{W}$, then

$$r(T) + n(T) = \dim \mathscr{V}.$$

PROOF  This is immediate from Theorem 3.3. Note that the dimension of $\mathscr{W}$ is immaterial. If the domain of a linear mapping is finite-dimensional, the dimension of the domain equals the sum of the rank and the nullity of the mapping.

We also note from Theorem 3.3 that the dimension of the range space of a linear mapping never exceeds the dimension of the domain space. Again a mapping diagram helps to keep the essential relations in mind, especially for successive mappings. Consider vector spaces $\mathscr{V}, \mathscr{W}$, and $\mathscr{X}$ and linear mappings $T$ from $\mathscr{V}$ to $\mathscr{W}$ and $S$ from $\mathscr{W}$ to $\mathscr{X}$. From Theorem 3.2, linear mappings carry subspaces into subspaces, and all vectors that are mapped into a subspace must comprise a subspace. In particular any vector that is mapped by $T$ into $\theta$ is also carried into $\theta$ by $ST$.

**THEOREM 3.5** If $T$ is a linear mapping from $\mathscr{V}$ to $\mathscr{W}$ and if $S$ is a linear mapping from $\mathscr{W}$ to $\mathscr{X}$, then

(a)  $\mathcal{N}(\mathbf{ST}) \supseteq \mathcal{N}(\mathbf{T})$ in $\mathcal{V}$,

(b)  $\mathcal{R}(\mathbf{ST}) \subseteq \mathcal{R}(\mathbf{S})$ in $\mathcal{X}$.

PROOF  Exercise. Show that if $\xi \in \mathcal{N}(\mathbf{T})$, then $\xi \in \mathcal{N}(\mathbf{ST})$, and if $\eta \in \mathcal{R}(\mathbf{ST})$, then $\eta \in \mathcal{R}(\mathbf{S})$. See Figure 3.3.

When $\mathbf{T}$ is a linear transformation on $\mathcal{V}$, the iteration of $\mathbf{T}$ produces a sequence of linear transformations: $\mathbf{T}$, $\mathbf{T}^2$, $\mathbf{T}^3$, and so on. From Theorem 3.5 we see that the null and range spaces of these transformations are nested subspaces of $\mathcal{V}$.

THEOREM 3.6  If $\mathbf{T}$ is a linear transformation on $\mathcal{V}_n$, then

(a)  $\mathcal{V}_n \supseteq \mathcal{R}(\mathbf{T}) \supseteq \mathcal{R}(\mathbf{T}^2) \supseteq \ldots \supseteq \mathcal{R}(\mathbf{T}^j) \supseteq \ldots$,

(b)  $[\theta] \subseteq \mathcal{N}(\mathbf{T}) \subseteq \mathcal{N}(\mathbf{T}^2) \subseteq \ldots \subseteq \mathcal{N}(\mathbf{T}^j) \subseteq \ldots$.

Furthermore, if $\mathcal{N}(\mathbf{T}^p) = \mathcal{N}(\mathbf{T}^{p+1})$, then

$$\mathcal{N}(\mathbf{T}^p) = \mathcal{N}(\mathbf{T}^{p+k}) \qquad \text{for all } k \geq 1,$$

$$\mathcal{R}(\mathbf{T}^p) = \mathcal{R}(\mathbf{T}^{p+k}),$$

$$\mathcal{V}_n = \mathcal{R}(\mathbf{T}^p) \oplus \mathcal{N}(\mathbf{T}^p).$$

PROOF  The chains are established by repeated application of Theorem 3.5. Since $\mathcal{V}_n$ is finite-dimensional, equality must hold somewhere in each chain. From Theorem 3.4 we know that $r(\mathbf{T}^i) + n(\mathbf{T}^i) = n$ for each $i$, so equality occurs at a given position in one chain if and only if it holds at the same position in the other. Suppose that $\mathcal{N}(\mathbf{T}^p) = \mathcal{N}(\mathbf{T}^{p+1})$, and let $\xi \in \mathcal{N}(\mathbf{T}^{p+k})$ for $k > 1$. Then $\theta = \mathbf{T}^{p+k}\xi = \mathbf{T}^{p+1}(\mathbf{T}^{k-1}\xi)$, so $\mathbf{T}^{k-1}\xi \in \mathcal{N}(\mathbf{T}^{p+1}) = \mathcal{N}(\mathbf{T}^p)$. Hence $\mathbf{T}^{p+k-1}\xi = \theta$, so $\xi \in \mathcal{N}(\mathbf{T}^{p+k-1})$. Therefore $\mathcal{N}(\mathbf{T}^{p+k}) \subseteq \mathcal{N}(\mathbf{T}^{p+k-1})$, and the opposite relation is given by (b), so equality must hold: $\mathcal{N}(\mathbf{T}^{p+k}) = \mathcal{N}(\mathbf{T}^{p+k-1})$ whenever $\mathcal{N}(\mathbf{T}^{p+1}) = \mathcal{N}(\mathbf{T}^p)$.

To prove that $\mathcal{V}_n$ is the direct sum of $\mathcal{R}(\mathbf{T}^p)$ and $\mathcal{N}(\mathbf{T}^p)$ we use Exercise 11 of Section 2.3. If $\eta \in \mathcal{R}(\mathbf{T}^p) \cap \mathcal{N}(\mathbf{T}^p)$, then $\eta = \mathbf{T}^p\xi$ for some $\xi \in \mathcal{V}_n$. And $\theta = \mathbf{T}^p\eta = \mathbf{T}^p(\mathbf{T}^p\xi)$, so $\xi \in \mathcal{N}(\mathbf{T}^{2p}) = \mathcal{N}(\mathbf{T}^p)$. Therefore $\theta = \mathbf{T}^p\xi = \eta$. Since the sum of the dimensions of $\mathcal{R}(\mathbf{T}^p)$ and $\mathcal{N}(\mathbf{T}^p)$ is $n$, $\mathcal{V}_n = \mathcal{R}(\mathbf{T}^p) \oplus \mathcal{N}(\mathbf{T}^p)$.

In summary, when the same linear transformation $\mathbf{T}$ is applied repeatedly to a finite-dimensional space $\mathcal{V}$, the null spaces $\mathcal{N}(\mathbf{T}^i)$ form a *strictly increasing* sequence of nested spaces until equality first occurs:

$\mathcal{N}(\mathbf{T}^p) = \mathcal{N}(\mathbf{T}^{p+1})$. Thereafter the null spaces do not change. The corresponding range spaces *strictly decrease* until $\mathcal{R}(\mathbf{T}^p) = \mathcal{R}(\mathbf{T}^{p+1})$. Also $\mathcal{V}$ is the direct sum of these two subspaces,

$$\mathcal{V} = \mathcal{R}(\mathbf{T}^p) \oplus \mathcal{N}(\mathbf{T}^p).$$

Finally **T** maps $\mathcal{R}(\mathbf{T}^p)$ *onto* itself, and $\mathbf{T}^p$ maps $\mathcal{N}(\mathbf{T}^p)$ onto $[\theta]$; if $p > 1$. **T** is nilpotent of index $p$ on $\mathcal{N}(\mathbf{T}^p)$.

## EXERCISES 3.2

1. Determine the rank and nullity of each of the linear transformations defined in Examples (c), (i), and (j) of Section 3.1. Verify the conclusions of Theorem 3.4 in each case.

2. Determine the range space and null space of the linear mappings **D** and **M** defined in Exercise 10 of Section 3.1.

3. Let **T** denote the linear transformation defined on $R_3$ by

$$\mathbf{T}(a_1, a_2, a_3) = (a_2, 0, 2a_3).$$

Carry out the calculations needed to verify all of the assertions of Theorem 3.6 for this particular example.

4. Show that if **T** is a linear transformation such that $r(\mathbf{T}) = 1$, then $\mathbf{T}^2 = c\mathbf{T}$ for some scalar $c$. Is the converse true? Explain.

5. Prove Theorem 3.5.

6. Let **S** and **T** be linear transformations on $\mathcal{V}_n$. Apply Theorems 3.5 and 3.4 to prove that
   (i) $n(\mathbf{ST}) \geq n(\mathbf{T})$ and $r(\mathbf{ST}) \leq r(\mathbf{S})$,
   (ii) $n(\mathbf{ST}) \geq n(\mathbf{S})$ and $r(\mathbf{ST}) \leq r(\mathbf{T})$.

7. Let **S** and **T** be linear transformations on $\mathcal{V}_n$. Prove that
   (i) $\mathcal{R}(\mathbf{S} + \mathbf{T}) \subseteq \mathcal{R}(\mathbf{S}) + \mathcal{R}(\mathbf{T})$,
   (ii) $r(\mathbf{S} + \mathbf{T}) \leq r(\mathbf{S}) + r(\mathbf{T})$,
   (iii) $n(\mathbf{S} + \mathbf{T}) \geq n(\mathbf{S}) + n(\mathbf{T}) - n$.

8. Let **S** and **T** be linear transformations on $\mathcal{V}_n$. Prove that
   (i) $n(\mathbf{S}) + n(\mathbf{T}) \geq n(\mathbf{ST}) \geq \max(n(\mathbf{S}), n(\mathbf{T}))$,
   (ii) $r(\mathbf{S}) + r(\mathbf{T}) - n \leq r(\mathbf{ST}) \leq \min(r(\mathbf{S}), r(\mathbf{T}))$.

9. Illustrate each of the assertions of Exercises 7 and 8 for the particular transformations **T** and **S** defined in Example (j) of Section 3.1.

## 3.3    Special Linear Transformations

In this section we single out four types of linear transformations for particular attention because of their importance in linear algebra and its applications; the four types of transformations are called *nonsingular*, *orthogonal*, *idempotent*, and *nilpotent*.

We shall consider the notion of nonsingularity in the more general context of a linear mapping from $\mathscr{V}$ to $\mathscr{W}$ instead of a transformation on $\mathscr{V}$. The basic concept of nonsingularity for linear mappings is the same as the idea of reversibility for arbitrary mappings, invertibility for real functions, and one-to-one-ness of algebraic homomorphisms. As we shall see, the range and null spaces are intrinsically related to any characterization of nonsingularity.

**DEFINITION 3.5**    A linear mapping **T** from $\mathscr{V}$ into $\mathscr{W}$ is said to be *nonsingular* if and only if there exists a mapping **S** from $\mathscr{R}(\mathbf{T})$ onto $\mathscr{V}$ such that **ST** is the identity mapping **I** on $\mathscr{V}$. Otherwise, **T** is said to be *singular*.

Observe that such an **S** must be linear, because for each $\gamma$, $\delta \in \mathscr{R}(\mathbf{T})$ there exist at least one $\alpha$ and one $\beta$ in $\mathscr{V}$ such that $\gamma = \mathbf{T}\alpha$, $\delta = \mathbf{T}\beta$. Since $\mathbf{ST} = \mathbf{I}$, we have $\mathbf{S}\gamma = \alpha$, $\mathbf{S}\delta = \beta$, and

$$\mathbf{S}(c\gamma + d\delta) = \mathbf{S}(c\mathbf{T}\alpha + d\mathbf{T}\beta) = \mathbf{ST}(c\alpha + d\beta)$$

$$= c\alpha + d\beta = c\mathbf{S}\gamma + d\mathbf{S}\delta.$$

**THEOREM 3.7**    The following statements are equivalent for a linear transformation **T** from $\mathscr{V}_n$ to $\mathscr{W}$.

(a)    **T** is nonsingular.

(b)    If $\mathbf{T}(\alpha) = \mathbf{T}(\beta)$ then $\alpha = \beta$.

(c)    $\mathscr{N}(\mathbf{T}) = [\theta]$.

(d)    $n(\mathbf{T}) = 0$.

(e)    $r(\mathbf{T}) = n$.

(f)    **T** maps any basis for $\mathscr{V}_n$ onto a basis for $\mathscr{R}(\mathbf{T})$.

**PROOF**    To say that the given statements are equivalent means that each statement implies and is implied by each of the others. Often it is possible to prove such equivalence by a cycle of implications:

$$\text{(a)} \Rightarrow \text{(b)} \Rightarrow \text{(c)} \Rightarrow \text{(d)} \Rightarrow \text{(e)} \Rightarrow \text{(f)} \Rightarrow \text{(a)}.$$

You may verify for yourself that all but the last of these implications are immediate consequences of the definitions and previous theorems. We now prove that (f) implies (a).

Let $\{\alpha_1, \ldots, \alpha_n\}$ be a basis for $\mathscr{V}_n$. Then by hypothesis $\{T\alpha_1, \ldots, T\alpha_n\}$ is a basis for $\mathscr{R}(T)$, and each $\eta \in \mathscr{R}(T)$ has a unique representation

$$\eta = \sum_{i=1}^{n} b_i T\alpha_i.$$

Let S be the mapping of $\mathscr{R}(T)$ into $\mathscr{V}_n$ defined by

$$S\eta = \sum_{i=1}^{n} b_i \alpha_i.$$

To show that $ST = I$ on $\mathscr{V}_n$, let $\xi \in \mathscr{V}_n$ have the representation

$$\xi = \sum_{i=1}^{n} a_i \alpha_i.$$

Then

$$T\xi = T\left(\sum_{i=1}^{n} a_i\alpha_i\right) = \sum_{i=1}^{n} a_i T\alpha_i.$$

By the definition of S,

$$S(T\xi) = S\left(\sum_{i=1}^{n} a_i T\alpha_i\right) = \sum_{i=1}^{n} a_i \alpha_i = \xi.$$

That is, $ST = I$ on $\mathscr{V}_n$.

As an exercise you may prove another useful property: T is non-singular if and only if T preserves the property of linear independence. In particular, Theorem 3.7 (f) assures us that any change of basis in $R_n$ is performed by a nonsingular linear transformation.

**THEOREM 3.8** Let T be a linear mapping from $\mathscr{V}$ to $\mathscr{W}$, and let S be a mapping from $\mathscr{R}(T)$ to $\mathscr{V}$ such that $ST = I$ on $\mathscr{V}$. Then $TS = I$ on $\mathscr{R}(T)$.

PROOF Since T is nonsingular, each $\gamma \in \mathscr{R}(T)$ is the T-image of one and only one $\alpha \in \mathscr{V}$: $\gamma = T\alpha$. Then $TS\gamma = TS(T\alpha) = T(ST)\alpha = TI\alpha = T\alpha = \gamma$, as desired.

This also shows that S is uniquely determined, since if $S_1T = I = S_2T$, then $S_2 = (S_1T)S_2 = S_1(TS_2) = S_1$, where Theorem 3.8 establishes the last equality. Hence when T is nonsingular and when $ST = I$ we shall call S the *inverse* of **T**, denoted by **T**⁻¹. Theorem 3.8 shows that if **ST** = **I**, then **TS** = **I**; observe, however, that the first **I** denotes the identity mapping from $\mathscr{V}$ to $\mathscr{V}$ and the second **I** denotes the identity mapping from $\mathscr{R}(\mathbf{T})$ to $\mathscr{R}(\mathbf{T})$.

If we specialize to the case of linear transformations on $\mathscr{V}_n$, then Theorem 3.7 (e) shows clearly that nonsingular transformations are precisely those which map $\mathscr{V}_n$ *onto* $\mathscr{V}_n$. A change of basis, for example, is nonsingular, and any nonsingular transformation is a change of basis.

More generally we can prove the following theorem.

**THEOREM 3.9**   Let **T** be a linear mapping from $\mathscr{V}_n$ to $\mathscr{W}_m$, and let **S** and **U** be nonsingular linear transformations of $\mathscr{V}_n$ and $\mathscr{W}_m$ respectively. Then
$$r(\mathbf{UTS}) = r(\mathbf{UT}) = r(\mathbf{TS}) = r(\mathbf{T}).$$

**PROOF**   The mapping **UTS** consists of a change of basis in $\mathscr{V}_n$, followed by a mapping **T** from $\mathscr{V}_n$ to $\mathscr{W}_m$, followed by a change of basis in $\mathscr{W}_m$. Since any change of basis preserves the dimension of all subspaces, the conclusion follows.

The second special type of transformation that we consider is an orthogonal linear transformation, which maps a Euclidean space onto itself in such a way that the length of every vector is preserved. Such transformations are rigid motions in which the origin remains fixed, and they frequently arise in physical and engineering applications of linear algebra.

**DEFINITION 3.6**   Let $\mathscr{V}$ be a Euclidean space. A linear transformation **T** on $\mathscr{V}$ is said to be *orthogonal* if and only if for every $\alpha \in \mathscr{V}$

$$\|\mathbf{T}\alpha\| = \|\alpha\|.$$

**THEOREM 3.10**   A linear transformation **T** on a Euclidean space $\mathscr{V}$ is orthogonal if and only if it preserves the inner product:

$$\langle \mathbf{T}\alpha, \mathbf{T}\beta \rangle = \langle \alpha, \beta \rangle \qquad \text{for all } \alpha, \beta \in \mathscr{V}.$$

**PROOF**   $\|\mathbf{T}(\alpha + \beta)\|^2 = \langle \mathbf{T}\alpha + \mathbf{T}\beta, \mathbf{T}\alpha + \mathbf{T}\beta \rangle$

$$= \langle \mathbf{T}\alpha, \mathbf{T}\alpha \rangle + 2\langle \mathbf{T}\alpha, \mathbf{T}\beta \rangle + \langle \mathbf{T}\beta, \mathbf{T}\beta \rangle$$

$$= \|\mathbf{T}\alpha\|^2 + 2\langle \mathbf{T}\alpha, \mathbf{T}\beta \rangle + \|\mathbf{T}\beta\|^2.$$

$$\|\alpha + \beta\|^2 = \langle \alpha + \beta, \ \alpha + \beta \rangle = \|\alpha\|^2 + 2\langle \alpha, \beta \rangle + \|\beta\|^2.$$

Comparing terms of these two equations and recalling that $\mathbf{T}$ preserves lengths, we obtain $\langle \mathbf{T}\alpha, \ \mathbf{T}\beta \rangle = \langle \alpha, \beta \rangle$. Conversely, if $\mathbf{T}$ preserves all inner products, it clearly preserves all lengths.

Hence we have proved that any length preserving transformation of Euclidean space also preserves the inner product and hence preserves angle and all metric properties. For this reason an orthogonal transformation is also called an *isometry*. We note that an orthogonal transformation is nonsingular since $\mathcal{N}(\mathbf{T}) = [\theta]$. Moreover, any linear transformation of $E_n$ that carries a normal orthogonal basis into a normal orthogonal basis is orthogonal.

The third special type of linear transformation discussed here was described in Definition 3.3. A nilpotent transformation $\mathbf{T}$ on $\mathcal{V}$ satisfies $\mathbf{T} \neq \mathbf{Z}$ but $\mathbf{T}^p = \mathbf{Z}$ for some integer $p > 1$. The derivative operator $\mathbf{D}$ is nilpotent on the polynomial space $\mathcal{P}_n$ but not on the space $\mathcal{P}$ of all real polynomials. Any nilpotent transformation is singular, but the converse is not true.

The remarks that follow the proof of Theorem 3.6 at the end of the previous section reveal the relation of nilpotent and nonsingular transformations to arbitrary transformations on a finite-dimensional space. For emphasis, we restate that result as a theorem:

**THEOREM 3.11** Let $\mathbf{T}$ be a linear transformation on $\mathcal{V}_n$. There exist two subspaces $\mathcal{M}$ and $\mathcal{N}$ such that

(a) $\mathcal{V}_n = \mathcal{M} \oplus \mathcal{N}$;

(b) $\mathcal{M}$ and $\mathcal{N}$ are $\mathbf{T}$-*invariant*; that is, $\mathbf{T}\mu \in \mathcal{M}$ for each $\mu \in \mathcal{M}$, and $\mathbf{T}\nu \in \mathcal{N}$ for each $\nu \in \mathcal{N}$;

(c) Restricted to $\mathcal{M}$, $\mathbf{T}$ is nonsingular;

(d) Restricted to $\mathcal{N}$, $\mathbf{T}$ is nilpotent.

The fourth type of linear transformation also was described in Definition 3.3. An idempotent linear transformation $\mathbf{T}$ on $\mathcal{V}$ satisfies $\mathbf{T}^2 = \mathbf{T}$. If $\mathbf{T}$ is idempotent and nonsingular, then $\mathbf{T}^{-1}(\mathbf{T}^2) = \mathbf{T}^{-1}\mathbf{T}$, so $\mathbf{T} = \mathbf{I}$. The geometric significance of idempotent transformations is revealed by the following theorem.

**THEOREM 3.12**    If $\mathbf{T}$ is an idempotent transformation on $\mathcal{V}_n$, then

(a)   $\mathcal{V}_n = \mathcal{R}(\mathbf{T}) \oplus \mathcal{N}(\mathbf{T})$;

(b)   $\mathcal{R}(\mathbf{T})$ and $\mathcal{N}(\mathbf{T})$ are $\mathbf{T}$-invariant;

(c)   For each $\eta \in \mathcal{R}(\mathbf{T})$, $\mathbf{T}(\eta) = \eta$;

(d)   For each $\nu \in \mathcal{N}(\mathbf{T})$, $\mathbf{T}(\nu) = \theta$.

PROOF    Theorem 3.6 with $p = 1$ yields (a) and (b), and (d) is trivial. If $\eta \in \mathcal{R}(\mathbf{T})$ then $\eta = \mathbf{T}\xi$ for some $\xi \in \mathcal{V}_n$; then $\mathbf{T}\eta = \mathbf{T}^2\xi = \mathbf{T}\xi = \eta$, since $\mathbf{T}^2 = \mathbf{T}$.

Hence an idempotent transformation on $\mathcal{V}_n$ is simply a *projection* of each point of $\mathcal{V}_n$ onto the subspace $\mathcal{R}(\mathbf{T})$. The essence of the Gram-Schmidt process is to obtain an orthogonal projection of a vector on a given subspace. Projections also provide the key idea for the *spectral theorem* — one of the most important results of linear algebra — which describes precisely those linear transformations that have a particularly simple and desirable form, namely, a linear combination of projections.

Other important properties of these special transformations are developed in the following exercises. You are urged to study them carefully.

### EXERCISES 3.3

1. Let $\mathbf{T}$ denote an arbitrary linear transformation of $R_2$, and suppose that $\mathbf{T}(1, 0) = (a, b)$ and $\mathbf{T}(0, 1) = (c, d)$. Show algebraically that $\mathbf{T}$ is nonsingular if and only if $ad - bc \neq 0$. Interpret geometrically.

2. Prove that if $\mathbf{T}$ is a nonsingular linear mapping, then $\mathbf{T}^{-1}$ is nonsingular, $(\mathbf{T}^{-1})^{-1} = \mathbf{T}$, and $(c\mathbf{T})^{-1} = c^{-1}\mathbf{T}^{-1}$ if $c \neq 0$.

3. Let $\mathbf{T}$ be a linear mapping from $\mathcal{V}$ *onto* $\mathcal{W}$, and $\mathbf{S}$ a linear mapping from $\mathcal{W}$ into $\mathcal{X}$. Prove that

(i)   $\mathbf{ST}$ is nonsingular if and only if both $\mathbf{T}$ and $\mathbf{S}$ are nonsingular;

(ii)   If $\mathbf{ST}$ is nonsingular, then $(\mathbf{ST})^{-1} = \mathbf{T}^{-1}\mathbf{S}^{-1}$.

4. Let $\mathbf{T}_1$ be a linear mapping of finite rank from $\mathcal{V}$ to $\mathcal{W}$, $\mathbf{T}$ a linear mapping from $\mathcal{R}(\mathbf{T}_1)$ into $\mathcal{X}$, and $\mathbf{T}_2$ a linear mapping from $\mathcal{R}(\mathbf{T})$ into $\mathcal{Y}$.

(i)   Draw a mapping diagram to show that $\mathbf{TT}_1$ and $\mathbf{T}_2\mathbf{T}$ are defined.

(ii)  As an extension of Exercise 6 of Section 3.2 prove that if **T** is non-singular, then $r(\mathbf{TT}_1) = r(\mathbf{T}_1)$ and $r(\mathbf{T}_2\mathbf{T}) = r(\mathbf{T}_2)$.

5. Prove that a linear mapping **T** from $\mathscr{V}_n$ to $\mathscr{W}$ is nonsingular if and only if for every linearly independent set $\{\alpha_1, \ldots, \alpha_k\}$ in $\mathscr{V}_n$ the set $\{\mathbf{T}\alpha_1, \ldots, \mathbf{T}\alpha_k\}$ is linearly independent in $\mathscr{W}$.

6. Given the space $\mathscr{P}$ of all real polynomials, let $\mathscr{P}_0$ be the subspace of all $p \in \mathscr{P}$ such that $p(0) = 0$. Consider the three linear mappings defined as follows:

$$\mathbf{J}p(x) = \int_0^x p(t)dt, \qquad \text{defined on } \mathscr{P},$$

$$\mathbf{D}p(x) = \frac{d}{dx}p(x), \qquad \text{defined on } \mathscr{P},$$

$$\mathbf{D}_0 p(x) = \frac{d}{dx}p(x), \qquad \text{defined on } \mathscr{P}_0.$$

(i)  Determine the domain and range of each of the seven mappings **J, D, D$_0$, DJ, JD, D$_0$J, JD$_0$**.

(ii)  Which of the seven mappings is the identity mapping on its domain?

(iii)  Which of the seven mappings is nonsingular?

7. Let **T** be an orthogonal transformation on a Euclidean space $E$. Suppose that there exist vectors $\alpha, \beta \in E$ and distinct scalars $a, b$ such that $\mathbf{T}\alpha = a\alpha$ and $\mathbf{T}\beta = b\beta$. Prove that $\alpha$ and $\beta$ are orthogonal.

8. Let **T** be a linear transformation which is nilpotent of index $p$ on $\mathscr{V}$. Show that there exists a **T**-invariant subspace $\mathscr{M}$ of dimension $p$. (See Exercise 11 of Section 3.1.)

9. Let **T** be a linear transformation which is nilpotent of index $p$ on $\mathscr{V}_n$. Show that for each $k < p$, $\mathscr{R}(\mathbf{T}^{p-k}) \subseteq \mathscr{N}(\mathbf{T}^k)$.

10. Show that a linear transformation **T** on $\mathscr{V}$ is idempotent if and only if $\mathbf{T}\eta = \eta$ for all $\eta \in \mathscr{R}(\mathbf{T})$.

11. Let $\mathscr{V}$ be the direct sum of two subspaces, $\mathscr{V} = \mathscr{M}_1 \oplus \mathscr{M}_2$. By Theorem 2.4, each $\xi \in \mathscr{V}$ has a unique representation $\xi = \mu_1 + \mu_2$ for some $\mu_1 \in \mathscr{M}_1, \mu_2 \in \mathscr{M}_2$. Let $\mathbf{E}_1, \mathbf{E}_2$ be the mappings defined on $\mathscr{V}$ by $\mathbf{E}_1\xi = \mu_1$, $\mathbf{E}_2\xi = \mu_2$.

(i)  Show that $\mathbf{E}_1$ and $\mathbf{E}_2$ are linear and idempotent. $\mathbf{E}_1$ is called the *projection on $\mathscr{M}_1$ along $\mathscr{M}_2$*, and $\mathbf{E}_2$ is the *projection on $\mathscr{M}_2$ along $\mathscr{M}_1$*.

(ii)  Show that $\mathbf{E}_1$ and $\mathbf{E}_2$ are *supplementary*, meaning that $\mathbf{E}_1 + \mathbf{E}_2 = \mathbf{I}$, and *orthogonal*, meaning that $\mathbf{E}_1\mathbf{E}_2 = \mathbf{E}_2\mathbf{E}_1 = \mathbf{Z}$.

(iii)  Illustrate geometrically in $R_3$.

## 3.4   Matrix Representation of a Linear Mapping

Up to this point in our study of linear mappings we have not placed much emphasis on particular coordinate systems for the vector spaces involved. Our major tools have been the concepts of mapping, linearity, and subspace, which are in essence geometric properties. Consequently our work has acquired a geometric, rather than algebraic or arithmetic, flavor. Coordinate-free methods should be used when they apply, not only because of their generality and simplicity, but also because we are most interested in discovering the *intrinsic* properties of the objects under study. By an intrinsic property we mean one that is valid in every coordinate system. For example, in the study of curves in the Euclidean plane a point of inflection is an intrinsic property of the curve; a relative maximum point is not, nor is the slope of a line.

In this section we shall see that a linear mapping from $\mathcal{V}_n$ to $\mathcal{W}_m$ can be represented algebraically or arithmetically by a set of numbers relative to a choice of coordinates, much in the same way that a vector can be represented. As in the study of analytic geometry, this algebraic representation can be exploited to discover additional geometric information, but in so doing we must be aware that the results obtained might depend upon the choice of bases.

Given a vector space $\mathcal{V}_n$, any choice of ordered basis provides a means of representing each vector uniquely as an ordered $n$-tuple of scalars. If $\xi = \sum_{j=1}^{n} x_j \alpha_j$, then the ordered $n$-tuple that represents $\xi$ relative to the $\alpha$-basis is $(x_1, \ldots, x_n)$. We could equally well write this $n$-tuple of scalars as a column

$$\begin{pmatrix} x_1 \\ \cdot \\ \cdot \\ \cdot \\ x_n \end{pmatrix}$$

rather than a row, and in fact we shall see that the column representation turns out to be quite convenient. Our immediate task, however, is to see that any linear mapping also can be represented by an array of scalars.

Let $\mathcal{V}_n$ and $\mathcal{W}_m$ be any two finite-dimensional vector spaces over a field $F$, and let $\mathbf{T}$ be a linear mapping from $\mathcal{V}_n$ to $\mathcal{W}_m$. We first choose an ordered basis $\{\alpha_1, \ldots, \alpha_n\}$ for $\mathcal{V}_n$ and an ordered basis $\{\beta_1, \ldots, \beta_m\}$

for $\mathscr{W}_m$. For any $\xi \in \mathscr{V}_n$, $\mathbf{T}\xi \in \mathscr{W}_m$ and hence $\mathbf{T}\xi$ has a unique representation as a linear combination of the $\beta_i$. Furthermore, linearity of $\mathbf{T}$ guarantees that if $\xi = \sum_{j=1}^{n} x_j \alpha_j$, then $\mathbf{T}\xi = \sum_{j=1}^{n} x_j \mathbf{T}\alpha_j$. Hence, if we know the effect of $\mathbf{T}$ on each vector of a basis for $\mathscr{V}_n$, we know the effect of $\mathbf{T}$ on every vector of $\mathscr{V}_n$. Therefore it is natural to examine $\mathbf{T}\alpha_j$ for each $j = 1, \ldots, n$:

$$\mathbf{T}\alpha_1 = a_{11}\beta_1 + a_{21}\beta_2 + \ldots + a_{m1}\beta_m,$$

$$\mathbf{T}\alpha_2 = a_{12}\beta_1 + a_{22}\beta_1 + \ldots + a_{m2}\beta_m,$$

$$\cdot \qquad \cdot \qquad \cdot \qquad \qquad \cdot$$
$$\cdot \qquad \cdot \qquad \cdot \qquad \qquad \cdot$$
$$\cdot \qquad \cdot \qquad \cdot \qquad \qquad \cdot$$

$$\mathbf{T}\alpha_n = a_{1n}\beta_1 + a_{2n}\beta_2 + \ldots + a_{mn}\beta_m.$$

In general, $\mathbf{T}\alpha_j = \sum_{i=1}^{m} a_{ij}\beta_i$, $j = 1, 2, \ldots, n$. Then, for

$$\xi = \sum_{j=1}^{n} x_j \alpha_j,$$

$$\mathbf{T}\xi = \mathbf{T}\left(\sum_{j=1}^{n} x_j \alpha_j\right) = \sum_{j=1}^{n} x_j \mathbf{T}\alpha_j$$

$$= \sum_{j=1}^{n} x_j \left(\sum_{i=1}^{m} a_{ij}\beta_i\right) = \sum_{i=1}^{m} \left(\sum_{j=1}^{n} x_j a_{ij}\right)\beta_i.$$

This expresses $\mathbf{T}\xi$ in terms of the scalars $x_j$ that represent $\xi$ relative to the $\alpha$-basis and the scalars $a_{ij}$ that represent $\mathbf{T}\alpha_j$ relative to the $\beta$-basis.

The *column* of scalars that represents the vector $\mathbf{T}\alpha_j$ is

$$\begin{pmatrix} a_{1j} \\ a_{2j} \\ \cdot \\ \cdot \\ \cdot \\ a_{mj} \end{pmatrix}, \qquad \text{for } j = 1, 2, \ldots, n.$$

If we write these columns in order, we obtain a rectangular array of scalars:

$$\begin{pmatrix} a_{11} & a_{12} & \cdots & a_{1n} \\ a_{21} & a_{22} & \cdots & a_{2n} \\ & & \cdot & \\ & & \cdot & \\ & & \cdot & \\ a_{m1} & a_{m2} & \cdots & a_{mn} \end{pmatrix}$$

**DEFINITION 3.7**  Any rectangular array of elements of a field $F$ is called a *matrix* over $F$. A matrix having $m$ rows and $n$ columns is called an *m-by-n* matrix. The scalar in row $i$ and column $j$ is denoted $a_{ij}$; the first subscript is the *row index*, and the second subscript is the *column index*. A one-by-$n$ matrix is a *row vector*. An $m$-by-one matrix is a *column vector*.

It is useful to summarize our conventions for representing a linear mapping $T$ from $\mathscr{V}_n$ to $\mathscr{W}_m$, relative to an ordered basis $\{\alpha_1, \ldots, \alpha_n\}$ for $\mathscr{V}_n$ and an ordered basis $\{\beta_1, \ldots, \beta_n\}$ for $\mathscr{W}_m$:

(a) $$\xi = \sum_{i=1}^n x_i\alpha_i \to X = \begin{pmatrix} x_1 \\ \cdot \\ \cdot \\ \cdot \\ x_n \end{pmatrix},$$

(b) $$T\alpha_j = \sum_{i=1}^m a_{ij}\beta_i \to C_j = \begin{pmatrix} a_{1j} \\ \cdot \\ \cdot \\ \cdot \\ a_{mj} \end{pmatrix}, j = 1, \ldots, n,$$

(c) $$T \to A = \begin{pmatrix} a_{11} & a_{12} & \cdots & a_{1n} \\ a_{21} & a_{22} & \cdots & a_{2n} \\ & \cdot & \cdot & \\ & \cdot & \cdot & \\ & \cdot & \cdot & \\ a_{m1} & a_{m2} & \cdots & a_{mn} \end{pmatrix}.$$

The entry in row $i$ and column $j$ is the coefficient $a_{ij}$ of $\beta_i$ in the expression for $T\alpha_j$. Therefore, the first column of $A$ represents $T\alpha_1$ relative to the $\beta$-basis; the second column of $A$ represents $T\alpha_2$, and so on.

(d)   $\eta = \mathbf{T}\xi = \sum_{i=1}^{m} y_i \beta_i \rightarrow Y = \begin{pmatrix} y_1 \\ \cdot \\ \cdot \\ \cdot \\ y_m \end{pmatrix}, \; y_i = \sum_{j=1}^{n} a_{ij} x_j.$

There is a simple way to remember how to calculate the scalars $y_i$ that represent $\eta = \mathbf{T}\xi$ in terms of the scalars $x_j$ that represent $\xi$ and the scalars $a_{ij}$ that represent $\mathbf{T}$. Imitating the equation $\eta = \mathbf{T}\xi$ we write $Y = AX$, or in extended form

$$\begin{pmatrix} y_1 \\ \cdot \\ \cdot \\ y_i \\ \cdot \\ \cdot \\ y_m \end{pmatrix} = \begin{pmatrix} a_{11} & \cdots & a_{1n} \\ \cdot & & \cdot \\ \cdot & & \cdot \\ a_{i1} & \cdots & a_{in} \\ \cdot & & \cdot \\ \cdot & & \cdot \\ a_{m1} & \cdots & a_{mn} \end{pmatrix} \begin{pmatrix} x_1 \\ \cdot \\ \cdot \\ \cdot \\ x_n \end{pmatrix},$$

and note that $y_i$ has the form of the standard scalar product in $E_n$ of the *i*th *row* vector of $A$ and the *column* vector $X$:

$$y_i = a_{i1}x_1 + a_{i2}x_2 + \ldots + a_{in}x_n.$$

This observation is a special case of a more general pattern which we shall now examine. Consider a third vector space, $\mathcal{X}_p$ with a chosen basis $\{\gamma_1, \ldots, \gamma_p\}$ and a linear mapping **S** from $\mathcal{W}_m$ to $\mathcal{X}_p$, as in Figure 3.4.

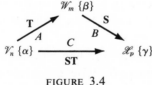

FIGURE 3.4

**T** is represented relative to the $\alpha$-basis and $\beta$-basis by an *m*-by-*n* matrix $A$. **S** is represented relative to the $\beta$-basis and $\gamma$-basis by a *p*-by-*m* matrix $B$. The composite mapping **ST** from $\mathcal{V}_n$ to $\mathcal{X}_p$ is represented relative to the $\alpha$-basis and $\gamma$-basis by a *p*-by-*n* matrix $C$. How is $C$ related to $A$ and $B$?

We have

$$T\alpha_j = \sum_{k=1}^{m} a_{kj}\beta_k, \; j = 1, \ldots, n;$$

$$S\beta_k = \sum_{i=1}^{p} b_{ik}\gamma_i, \; k = 1, \ldots, m.$$

Then

$$(ST)\alpha_j = S(T\alpha_j) = S\left(\sum_{k=1}^{m} a_{kj}\beta_k\right) = \sum_{k=1}^{m} a_{kj}S\beta_k$$

$$= \sum_{k=1}^{m} a_{kj}\left(\sum_{i=1}^{p} b_{ik}\gamma_i\right) = \sum_{i=1}^{p} \left(\sum_{k=1}^{m} b_{ik}a_{kj}\right)\gamma_i.$$

This tells us that in the matrix $C$ the entry in row $i$ and column $j$ (that is, the scalar coefficient $c_{ij}$ of $\gamma_i$ in $ST\alpha_j$) is given by

$$c_{ij} = \sum_{k=1}^{m} b_{ik}a_{kj}.$$

Thus $c_{ij}$ is the standard scalar product in $\mathscr{W}_m$ of the $i$th row vector of $B$ with the $j$th column vector of $A$, for each $i = 1, 2, \ldots, p$ and each $j = 1, 2, \ldots, n$.

It is precisely this computation that motivates the definition of the product of matrices, given in the next section. Our object is to devise an algebraic system for matrices that mirrors the algebra of linear mappings. If the matrix $A$ represents the mapping $T$, and if the matrix $B$ represents the mapping $S$, we want to define matrix multiplication so that the matrix $BA$ represents the mapping $ST$. And if the vector $\xi$ is represented by $X$ we want $AX$ to represent $T\xi$. The reason we chose to represent a vector by a column of scalars instead of a row of scalars was simply to make the matrix algebra work out nicely as an imitation of the algebra of linear mappings.

To illustrate, we consider the linear transformation $T_\Psi$ that rotates the Euclidean plane counterclockwise around the origin through an angle $\Psi$, as in Example (c) of Section 3.1. Relative to $\{\epsilon_1, \epsilon_2\}$ the matrix for $T_\Psi$ is

$$A = \begin{pmatrix} \cos \Psi & -\sin \Psi \\ \sin \Psi & \cos \Psi \end{pmatrix}.$$

For an arbitrary point $(x,y)$, $T_\Psi(x,y)$ can be calculated as

$$\begin{pmatrix} \cos \Psi & -\sin \Psi \\ \sin \Psi & \cos \Psi \end{pmatrix} \begin{pmatrix} x \\ y \end{pmatrix} = \begin{pmatrix} x' \\ y' \end{pmatrix},$$

where $x' = x\cos \Psi - y\sin \Psi$, $y' = x\sin \Psi + y\cos \Psi$. Furthermore the effect of successive rotation $T_\omega T_\Psi$, first through $\Psi$ and then through $\omega$, can be computed from the matrices according to the formula $c_{kj} = \sum_{i=1}^{2} b_{ki}a_{ij}$:

$$\begin{pmatrix} \cos \omega & -\sin \omega \\ \sin \omega & \cos \omega \end{pmatrix} \begin{pmatrix} \cos \Psi & -\sin \Psi \\ \sin \Psi & \cos \Psi \end{pmatrix} = \begin{pmatrix} \cos(\omega + \Psi) & -\sin(\omega + \Psi) \\ \sin(\omega + \Psi) & \cos(\omega + \Psi) \end{pmatrix}.$$

The computations needed to verify the entries of the matrix on the right are illustrated as follows. For the entry $c_{11}$ in the first row and first column we compute the dot product of the first row of $A_\omega$ and the first column of $A_\Psi$:

$$(\cos \omega)(\cos \Psi) + (-\sin \omega)(\sin \Psi) = \cos(\omega + \Psi).$$

For $c_{12}$ we have

$$(\cos \omega)(-\sin \Psi) + (-\sin \omega)(\cos \Psi) = -\sin(\omega + \Psi).$$

The entries $c_{21}$ and $c_{22}$ are calculated in the same way.

## EXERCISES 3.4

1. A linear transformation $T$ of $R_2$ is defined by $T\alpha_1 = \alpha_1 - \alpha_2$, $T\alpha_2 = \alpha_1 + \alpha_2$, where $\alpha_1 = \epsilon_1 + 2\epsilon_2$ and $\alpha_2 = \epsilon_1 - \epsilon_2$.

   (i) Write the matrix $A$ for $T$ relative to the $\alpha$-basis.

   (ii) Write the matrix $B$ for $T$ relative to the $\epsilon$-basis.

   (iii) If $\xi = x\epsilon_1 + y\epsilon_2$, write a column vector $x$ for $\xi$ relative to the $\epsilon$-basis and a column vector $Y$ for $\xi$ relative to the $\alpha$-basis.

   (iv) Calculate $BX$ and show that it represents $T\xi$ relative to the $\epsilon$-basis. Show also that $AY$ represents $T\xi$ relative to the $\alpha$-basis.

   (v) Check your results by computing the $\epsilon$-components of the vector whose $\alpha$-components are given by $AY$.

2. Referring to Example (d) of Section 1.10, confirm the results expressed there by using matrix computations.

   (i) Write the matrix $A$ for $T$ relative to the $\epsilon$-basis.

   (ii) Write the matrix $B$ for $T$ relative to the $\alpha$-basis.

   (iii) Compute $T\alpha_1$, $T\alpha_2$, $T\alpha_3$ using the $\epsilon$-representations.

   (iv) Compute $T\xi$ using the $\alpha$-representations.

3. A linear transformation $T$ of the Euclidean plane carries $(1, 1)$ into $(-2, 0)$ and also carries $(0, 1)$ into $(-1, 1)$.

   (i)   Write the matrix $A$ for $T$ relative to $\{\epsilon_1, \epsilon_2\}$.

   (ii)  Let $\gamma_1 = (1, 1)$ and $\gamma_2 = (0, 1)$. Write the matrix $C$ for $T$ relative to $\{\gamma_1, \gamma_2\}$.

   (iii) If $\eta = -\epsilon_1 + \epsilon_2$, write column vectors $X$ and $Y$ for $\eta$ relative to $\{\epsilon_1, \epsilon_2\}$ and $\{\gamma_1, \gamma_2\}$ respectively.

   (iv)  Compute $AX$ and $CY$, and show that each represents $T^2\gamma_2$.

4. In $R_2$ let $\alpha_1 = \epsilon_1 + 2\epsilon_2$, $\alpha_2 = -\epsilon_1 + \epsilon_2$, and let $S$ be the linear transformation which carries out the change of basis: $S\epsilon_1 = \alpha_1$, $S\epsilon_2 = \alpha_2$.

   (i)   Write the matrix $B$ for $S$ relative to $\{\epsilon_1, \epsilon_2\}$.

   (ii)  Write the matrix $D$ for $S$ relative to $\{\alpha_1, \alpha_2\}$.

   (iii) Write the matrix $E$ for $S^{-1}$ relative to $\{\epsilon_1, \epsilon_2\}$.

   (iv)  Compute $BE$ and $EB$.

   (v)   Compute $S(4\alpha_1 - 3\alpha_2)$ in three ways: directly from the definitions of $S$, $\alpha_1$, and $\alpha_2$; by matrix calculations in the $\epsilon$-basis; and by matrix calculations in the $\alpha$-basis. Reconcile the results.

5. Let $S$ and $T$ be the transformations defined in Exercises 3 and 4, represented by matrices $B$ and $A$ relative to $\{\epsilon_1, \epsilon_2\}$.

   (i)   By computing $ST\epsilon_1$ and $ST\epsilon_2$, write a matrix $F$ for $ST$ relative to $\{\epsilon_1, \epsilon_2\}$.

   (ii)  Check your results by applying the formula $f_{ij} = \sum_{k=1}^{2} b_{ik}a_{kj}$.

   (iii) Perform the computations of (i) and (ii) for $TS$.

   (iv)  Calculate the image of an arbitrary point $(x, y)$ under each of the transformations $ST$ and $TS$.

6. Let $T$ be idempotent on $\mathscr{V}_n$. According to Theorem 3.12, if $B_1$ is a basis for $\mathscr{R}(T)$ and if $B_2$ is a basis for $\mathscr{N}(T)$, then $B_1 \cup B_2$ is a basis for $\mathscr{V}_n$. What matrix represents $T$ relative to the basis $B_1 \cup B_2$?

7. Let $T$ be nilpotent of index $n$ on $\mathscr{V}_n$. According to Exercise 11 of Section 3.1, if $T^{n-1}\xi \neq \theta$, then $\{\xi, T\xi, \ldots, T^{n-1}\xi\}$ is a basis for $\mathscr{V}_n$. What matrix represents $T$ relative to this basis?

8. Given two linear mappings $T$ and $S$ from $\mathscr{V}_n$ to $\mathscr{W}_m$, let $A = (a_{ij})$ and $B = (b_{ij})$ respectively be the $m$-by-$n$ matrices that represent $T$ and $S$, relative to a basis $\{\alpha_1, \ldots, \alpha_n\}$ for $\mathscr{V}_n$ and a basis $\{\beta_1, \ldots, \beta_m\}$ for $\mathscr{W}_m$.

(i) Show that the matrix $C = (c_{ij})$ that represents $\mathbf{T} + \mathbf{S}$ is given by $c_{ij} = a_{ij} + b_{ij}$ for $i = 1, \ldots, m;\ j = 1, \ldots, n$.

(ii) Show that the matrix $D = (d_{ij})$ that represents $k\mathbf{T}$ is given by $d_{ij} = ka_{ij}$ for $i = 1, \ldots, m;\ j = 1, \ldots, n$.

## 3.5 Matrix Algebra

Since linear mappings are represented by matrices our sole guide in defining algebraic operations for matrices is to guarantee that the algebra of linear mappings is represented in all respects by the algebra of matrices. We recall a few facts about the algebra of linear transformations. If $\mathbf{T}$ and $\mathbf{S}$ are linear mappings from $\mathscr{V}_n$ to $\mathscr{W}_m$ and if $k$ is any scalar, then $\mathbf{T} + \mathbf{S}$ and $k\mathbf{T}$ are linear mappings from $\mathscr{V}_n$ to $\mathscr{W}_m$:

$$\mathbf{T} = \mathbf{S} \text{ if and only if } \mathbf{T}\xi = \mathbf{S}\xi \text{ for each } \xi \in \mathscr{V}_n,$$

$$(\mathbf{T} + \mathbf{S})\xi = \mathbf{T}\xi + \mathbf{S}\xi,$$

$$(k\mathbf{T})\xi = k(\mathbf{T}\xi).$$

Since equality, addition, and scalar multiples of vectors are defined component by component, the corresponding operations for matrices are defined in the same way.

**DEFINITION 3.8** Let $A = (a_{ij})$ and $B = (b_{ij})$ be $m$-by-$n$ matrices and let $k$ be a scalar. Matrix *equality*, *sum*, and *scalar multiple* are defined as follows:

(a) $A = B$ if and only if $a_{ij} = b_{ij}$ for each $i = 1, \ldots, m$ and each $j = 1, \ldots, n$;

(b) $A + B = (c_{ij})$, where $c_{ij} = a_{ij} + b_{ij}$ for each $i = 1, \ldots, m$ and each $j = 1, \ldots, n$;

(c) $kA = (d_{ij})$, where $d_{ij} = ka_{ij}$ for each $i = 1, \ldots, m$ and each $j = 1, \ldots, n$.

To illustrate, if

$$A = \begin{pmatrix} 2 & -1 & -2 \\ 1 & -3 & -1 \end{pmatrix} \text{ and } B = \begin{pmatrix} -1 & 3 & 0 \\ 4 & 2 & 1 \end{pmatrix},$$

then

$$A + B = \begin{pmatrix} 1 & 2 & -2 \\ 5 & -1 & 0 \end{pmatrix} \text{ and } 3B = \begin{pmatrix} -3 & 9 & 0 \\ 12 & 6 & 3 \end{pmatrix}.$$

**THEOREM 3.13**    The system $\mathscr{M}_{m \times n}$ of all $m$-by-$n$ matrices of elements from a field $F$ forms a vector space of dimension $mn$ over $F$, relative to the operations of matrix sum and scalar multiple.

**PROOF**    It is trivial to verify that $\mathscr{M}_{m \times n}$ satisfies the definition of a vector space. The $m$-by-$n$ matrix $Z$ having 0 as each entry is the zero matrix, and $-1A$ is the additive inverse of $A$. To obtain a basis for $\mathscr{M}_{m \times n}$ consider the set of $mn$ distinct matrices $U_{ij}, i = 1, \ldots, m$ and $j = 1, \ldots, n$, where $U_{ij}$ has 1 in the $(i, j)$ position and 0 in each of the other positions. For any $mn$ scalars $c_{ij}$, $\sum\limits_{i=1}^{m} \sum\limits_{j=1}^{n} c_{ij} U_{ij}$ is the matrix $C = (c_{ij})$. The $U_{ij}$ span $\mathscr{M}_{m \times n}$ and are linearly independent.

**THEOREM 3.14**    Let $\{\alpha_1, \ldots, \alpha_n\}$ be a basis for $\mathscr{V}_n$ and $\{\beta_1, \ldots, \beta_m\}$ a basis for $\mathscr{W}_m$. Each $m$-by-$n$ matrix $A = (a_{ij})$ determines a unique linear mapping $\mathbf{T}_A$ from $\mathscr{V}_n$ to $\mathscr{W}_m$, defined by

$$\mathbf{T}_A(\alpha_j) = \sum_{i=1}^{m} a_{ij} \beta_i.$$

Then the correspondence $A \to \mathbf{T}_A$ is an isomorphism of the vector space $\mathscr{M}_{m \times n}$ of all $m$-by-$n$ matrices onto $\mathscr{L}(\mathscr{V}_n, \mathscr{W}_m)$.

**PROOF**    The correspondence $A \to \mathbf{T}_A$ is a mapping from $\mathscr{M}_{m \times n}$ onto $\mathscr{L}(\mathscr{V}_n, \mathscr{W}_m)$ since, relative to the chosen bases, each $\mathbf{T}$ determines a matrix $B$ such that $\mathbf{T} = \mathbf{T}_B$. Furthermore the mapping is one-to-one, because if $A \neq B$ then $A$ and $B$ must differ in some column, say column $k$, and thus

$$\mathbf{T}_A(\alpha_k) \neq \mathbf{T}_B(\alpha_k).$$

The definition of matrix sum and scalar multiple then guarantees that $\mathbf{T}_{A+B} = \mathbf{T}_A + \mathbf{T}_B$ and $\mathbf{T}_{cA} = c\mathbf{T}_A$, because

$$\mathbf{T}_{A+B}(\alpha_j) = \sum_{i=1}^{m} (a_{ij} + b_{ij})\beta_i = \sum_{i=1}^{m} a_{ij}\beta_i + \sum_{i=1}^{m} b_{ij}\beta_i$$

$$= \mathbf{T}_A(\alpha_j) + \mathbf{T}_B(\alpha_j),$$

and

$$\mathbf{T}_{cA}(\alpha_j) = \sum_{i=1}^{m} (ca_{ij})\beta_i = c \sum_{i=1}^{m} a_{ij}\beta_i = c\mathbf{T}_A(\alpha_j).$$

The mapping $A \to \mathbf{T}_A$ is therefore a vector space isomorphism.

As a corollary of the last two theorems we now know that the dimension of the vector space $\mathscr{L}(\mathscr{V}_n, \mathscr{W}_m)$ is $mn$. This result is an example of one that seems to be easier to prove in terms of matrices than in terms of linear mappings. However, other results about matrices will be easier to prove in the form of linear mappings. This fundamental isomorphism theorem therefore provides a very powerful tool for the study of linear algebra. We now have two identical systems — linear mappings and matrices. The former is geometric, the latter numerical. But any theorem proved in one system can be translated into a theorem of the other system, with no further proof required.

It is important to observe, however, that the isomorphism of Theorem 3.14 was defined in terms of *fixed* bases for $\mathscr{V}_n$ and $\mathscr{W}_m$. Ordinarily the matrix $A$ that represents $T$ relative to one pair of bases will be *different* from the matrix $B$ that represents $T$ relative to another pair. Or the other way around, a matrix $A$ determines one linear mapping $T_A$ relative to one pair of bases but a *different* linear mapping $S_A$ relative to another pair.

Next consider how matrix multiplication can be defined in order to simulate the composition of linear mappings. The geometric framework within which the composition of mappings can occur was shown in Figure 3.4 of Section 3.4; if $T$ maps $\mathscr{V}_n$ into $\mathscr{W}_m$ and $S$ maps $\mathscr{W}_m$ into $\mathscr{X}_p$, then $ST$ is defined from $\mathscr{V}_n$ into $\mathscr{X}_p$. $S$ is represented by a $p$-by-$m$ matrix, $T$ is represented by an $m$-by-$n$ matrix, and the product $ST$ is represented by a $p$-by-$n$ matrix.

**DEFINITION 3.9** If $B = (b_{ik})$ is a $p$-by-$m$ matrix and if $A = (a_{kj})$ is an $m$-by-$n$ matrix, then the *product* $BA$ is the $p$-by-$n$ matrix $C = (c_{ij})$, where

$$c_{ij} = \sum_{k=1}^{m} b_{ik}a_{kj},$$

for each $i = 1, \ldots, p$ and each $j = 1, \ldots, n$.

Again we recognize the form of $c_{ij}$ as that of the standard scalar product in $E_m$ of row $i$ of $B$ with column $j$ of $A$. Again from this observation it is evident that a matrix product is defined only when the number of columns of the first matrix coincides with the number of rows of the second.

where $c_{ij} = b_{i1}a_{1j} + b_{i2}a_{2j} + \ldots + b_{im}a_{mj}$.

For example, if $A$ is three-by-two and $B$ is two-by-three, then both $AB$ and $BA$ are defined, but not equal nor even of the same size.

$$A = \begin{pmatrix} 2 & -1 \\ 1 & 0 \\ 0 & 3 \end{pmatrix}, \; B = \begin{pmatrix} 1 & -1 & 0 \\ 0 & -2 & 3 \end{pmatrix}$$

$$AB = \begin{pmatrix} 2 & 0 & -3 \\ 1 & -1 & 0 \\ 0 & -6 & 9 \end{pmatrix}, \; BA = \begin{pmatrix} 1 & -1 \\ -2 & 9 \end{pmatrix}$$

We now see that Theorem 3.14 can be strengthened in the special case of linear transformations of a single space $\mathscr{V}_n$ into itself. Then only one basis is needed, $\{\alpha_1, \ldots, \alpha_n\}$. The vector space $\mathscr{L}(\mathscr{V}_n, \mathscr{V}_n)$ is isomorphic to the space $\mathscr{M}_{n \times n}$, but as we saw in Section 3.1, $\mathscr{L}$ is also a linear algebra in which an associative, bilinear product operation is defined for linear transformations, namely the composition or successive application of transformations. Furthermore, matrix multiplication was defined expressly to imitate the composition of linear mappings. If $B$ represents $\mathbf{S}$ and $A$ represents $\mathbf{T}$, then $BA$ represents $\mathbf{ST}$. Hence the correspondence $A \rightarrow \mathbf{T}_A$ preserves the product operation as well as sum and scalar multiple.

**THEOREM 3.15**    The linear algebra $\mathscr{M}_{n \times n}$ of all $n$-by-$n$ matrices over $F$ is isomorphic to the linear algebra $\mathscr{L}(\mathscr{V}_n, \mathscr{V}_n)$ of all linear transformations on $\mathscr{V}_n$, where $\mathscr{V}_n$ is any $n$-dimensional vector space over $F$.

Observe that we did not prove directly that the product of matrices is associative and bilinear. Such proofs are straightforward, but the notation is tedious. Instead, we combined some knowledge of linear transformations with the fact that the isomorphism $A \rightarrow \mathbf{T}_A$ also preserves the product operation: $\mathbf{T}_{BA} = \mathbf{ST} = \mathbf{T}_B \mathbf{T}_A$. By appealing to Theorem 3.15 and previously proved properties of linear mappings we can therefore obtain the following properties of matrix algebra.

**THEOREM 3.16**    Let $A$, $B$, $C$ be $m$-by-$n$ matrices, $Z$ the $m$-by-$n$ matrix of zeros, and $a$, $b$ scalars. Then

(a)  $A + B = B + A$,

(b)  $(A + B) + C = A + (B + C)$,

(c)  $(a + b)A = aA + bA$,

(d) $a(A + B) = aA + aB,$

(e) $(ab)A = a(bA),$

(f) $A + Z = A,$

(g) $A + (-1)A = Z,$

(h) $0A = Z = aZ.$

**THEOREM 3.17** Let $A, B, C$ be $n$-by-$n$ matrices and $a, b$ scalars, then

(a) $C(BA) = (CB)A,$

(b) $C(aA + bB) = aCA + bCB,$

(c) $(aA + bB)C = aAC + bBC.$

These properties of matrix products hold also for rectangular matrices, provided of course that the dimensions of the matrices are such that all of the indicated sums and products are defined.

## EXERCISES 3.5

1. Compute $AB, AC, B^2, BC, CA$, given that

$$A = \begin{pmatrix} 1 & 0 & -1 \\ 0 & 2 & 3 \end{pmatrix}, B = \begin{pmatrix} 2 & -1 & 4 \\ 1 & 0 & -2 \\ 0 & 3 & 1 \end{pmatrix}, C = \begin{pmatrix} 0 & 2 \\ -1 & 0 \\ 3 & 1 \end{pmatrix}.$$

Are any other binary products possible for these three matrices?

2. Let $I_m = (\delta_{ij})$ denote the $m$-by-$m$ matrix with $\delta_{ij} = 1$ if $i = j$ and $\delta_{ij} = 0$ if $i \neq j$; hence $\delta_{ij}$ is the Kronecker delta. Prove that if $A$ is any $m$-by-$n$ matrix, then $I_m A = A$ and $A I_n = A$.

3. Given the matrices

$$A = \begin{pmatrix} 1 & 0 \\ 0 & -1 \end{pmatrix}, B = \begin{pmatrix} 0 & 1 \\ 1 & 0 \end{pmatrix}, C = \begin{pmatrix} 1 & 0 \\ 2 & 0 \end{pmatrix}.$$

(i) Describe geometrically the linear transformation of $R_2$ which each matrix represents relative to the standard $\epsilon$-basis.

(ii) Calculate the matrix products $AB, BA, A^2, B^2, C^2$ and interpret each as a geometric transformation.

4. Let $\{\alpha_1, \alpha_2, \alpha_3\}$ be any basis for $R_3$ and let

$$\begin{aligned} \beta_1 &= \alpha_1 - 2\alpha_2 \quad , \\ \beta_2 &= \alpha_1 + \alpha_2 + \alpha_3, \\ \beta_3 &= \alpha_2 - \alpha_3. \end{aligned}$$

(i)  Prove that $\{\beta_1, \beta_2, \beta_3\}$ is a basis, and express each $\alpha_i$ as a linear combination of the $\beta_j$.

(ii)  If **T** is defined by $\mathbf{T}(\alpha_i) = \beta_i$ for $i = 1, 2, 3$, find the matrix $A$ that represents **T** relative to the $\alpha$-basis.

(iii)  If **S** is defined by $\mathbf{S}(\beta_i) = \alpha_i$ for $i = 1, 2, 3$, find the matrix $B$ that represents **S** relative to the $\beta$-basis.

(iv)  By matrix computations show that $BA = I_3 = AB$.

5.  Represent each of the transformations **T** and **S** defined in Example (j) of Section 3.1 by a matrix and use matrix computations to verify each of the results stated there.

6.  Show that the system of all one-by-one matrices over a field $F$, together with matrix addition and multiplication, is a field that is isomorphic to $F$.

7.  Prove that the set of all real two-by-two matrices of the form

$$\begin{pmatrix} a & b \\ -b & a \end{pmatrix}$$

forms a system that is isomorphic to the field of complex numbers.

8.  From Exercise 7 of Section 3.4 we might expect the matrix

$$N = \begin{pmatrix} 0 & 0 & 0 & 0 \\ 1 & 0 & 0 & 0 \\ 0 & 1 & 0 & 0 \\ 0 & 0 & 1 & 0 \end{pmatrix}$$

to be nilpotent of index 4. Use matrix computation to show that it is.

9.  From Exercise 6 of Section 3.4 we might expect the matrix

$$A = \begin{pmatrix} 1 & 0 & 0 & 0 \\ 0 & 1 & 0 & 0 \\ 0 & 0 & 0 & 0 \\ 0 & 0 & 0 & 0 \end{pmatrix}$$

to be idempotent. Use matrix computation to show that it is.

10.  (i)  Use the isomorphism theorem and Theorem 3.8 to show that if $A$ and $X$ are $n$-by-$n$ matrices such that $XA = I_n$, then $AX = I_n$ where $I_n$ is defined as in Exercise 2.

(ii)  Specialize to the case where $m = n = 2$, and let $A = (a_{ij})$ and $X = (x_{jk})$. Write a system of four linear equations in the $a$'s and $x$'s that is equivalent to the matrix equation $XA = I_2$. Also write a system of four linear equations in the $a$'s and $x$'s that is equivalent to the matrix equation $AX = I_2$. Note that it is not entirely obvious that the $x$'s that solve the first system also solve the second.

11. In the vector space $\mathscr{P}_n$ of all real polynomials of degree not exceeding $n$, let $B$ denote the basis $\{1, x, \ldots, x^n\}$ and let $T_1$ and $T_2$ denote the transformation on $\mathscr{P}_n$ defined by

$$T_1(p(x)) = xp'(x),$$
$$T_2(p(x)) = x^2p''(x),$$

where $p'$ and $p''$ are the first and second derivatives of the polynomial $p$.

(i)  Prove that $T_1$ and $T_2$ are linear.

(ii)  Describe the null space and the range space of $T_2$, and thus determine the nullity and rank of $T_2$.

(iii)  Relative to the basis $B$ represent $T_1$ and $T_2$ by matrices $A_1$ and $A_2$.

(iv)  Prove directly that $T_1{}^2 = T_1 + T_2$.

(v)  Verify by matrix calculations that $A_1{}^2 = A_1 + A_2$.

## 3.6    Special Matrices

In this section we shall consider various types of *square* matrices that are important in a general study of matrix algebra either because they have particularly simple computational properties or because they represent special types of linear transformations.

In any matrix $A = (a_{ij})$, the positions for which $i = j$ comprise the *main diagonal* of $A$, and the scalars $a_{ii}$ are called *diagonal* entries. The scalars $a_{i+1,i}$ that appear in $A$ immediately below the diagonal elements are called *subdiagonal* entries.

*Identity Matrix.*    The $n$-by-$n$ matrix $I_n$ having 1 in each diagonal position and 0 in every other position is called the $n$-by-$n$ *identity matrix*:

$$I_n = \begin{pmatrix} 1 & 0 & . & . & 0 \\ 0 & 1 & . & . & 0 \\ . & . & & & . \\ . & . & & & . \\ 0 & 0 & . & . & 1 \end{pmatrix} = (\delta_{ij}), \qquad \delta_{ij} = \begin{cases} 1 \text{ if } i = j \\ 0 \text{ if } i \neq j \end{cases}.$$

By direct computation you may verify that

$$\text{if } A \text{ is } m\text{-by-}n, \text{ then } AI_n = A,$$

$$\text{if } B \text{ is } n\text{-by-}p, \text{ then } I_nB = B.$$

Hence $I_n$ is the multiplicative identity for the set of all $n$-by-$n$ matrices. Usually there is no need to emphasize the size of $I_n$, and the subscript $n$ can be omitted.

*Scalar Matrix.* Any scalar multiple of the $n$-by-$n$ identity matrix is called a *scalar matrix*:

$$K = kI = \begin{pmatrix} k & 0 & . & . & 0 \\ 0 & k & . & . & 0 \\ . & . & & & . \\ . & . & & & . \\ 0 & 0 & . & . & k \end{pmatrix}.$$

Clearly a scalar $n$-by-$n$ matrix commutes with *every* $n$-by-$n$ matrix $A$; that is, $KA = AK$. As an exercise you may show that any matrix that commutes with every $n$-by-$n$ matrix must be scalar. Furthermore, the mapping $k \to kI$ is a field isomorphism from $F$ onto the set of all $n$-by-$n$ scalar matrices.

*Diagonal Matrix.* A square matrix $D$ in which each nondiagonal entry is 0 is called a *diagonal matrix*:

$$D = \begin{pmatrix} a_{11} & 0 & . & . & 0 \\ 0 & a_{22} & . & . & 0 \\ . & . & & & . \\ . & . & & & . \\ 0 & 0 & . & . & a_{nn} \end{pmatrix} = \operatorname{diag}(a_{11}, \ldots, a_{nn}).$$

The sum, scalar multiple, and product of diagonal matrices are diagonal, and all $n$-by-$n$ diagonal matrices commute with each other. These facts make diagonal matrices especially easy to work with, and a major problem of linear algebra is to determine whether a given linear transformation can be represented by a diagonal matrix by a proper choice of basis.

*Triangular Matrix.* A square matrix in which every element *below* the main diagonal is zero is called *upper triangular*; $A = (a_{ij})$ where $a_{ij} = 0$ whenever $i > j$. A *lower triangular* matrix $B = (b_{ij})$ satisfies $b_{ij} = 0$ whenever $i < j$. The sum, scalar multiple, and product of lower triangular matrices is lower triangular; the corresponding statement is true for upper triangular matrices. A lower triangular matrix in which each diagonal element is zero is called *strictly* lower triangular; $C = (c_{ij})$ where

$c_{ij} = 0$ whenever $i \leq j$. Using more advanced methods one can prove that any linear transformation on Euclidean $n$-space can be represented by a lower triangular matrix relative to a normal orthogonal basis. This result is related to the problem of diagonal representation stated at the end of the previous paragraph. We also recall that the process of Gaussian elimination replaces a system of $m$ linear equations in $n$ unknowns by an equivalent system in echelon form. If $m = n$, the corresponding matrix is upper triangular.

*Diagonal Block Matrix.*  As a generalization of diagonal matrices, a *diagonal block matrix* is a square matrix of the form

$$B = \begin{pmatrix} B_1 & Z & . & . & Z \\ Z & B_2 & . & . & Z \\ . & . & & & . \\ . & . & & & . \\ Z & Z & . & . & B_k \end{pmatrix},$$

where each $B_i$ is a square matrix and each $Z$ represents a matrix of zeros. $B$ is diagonal if each $B_j$ is one-by-one. Such matrices arise whenever $\mathscr{V}_n$ can be written as a direct sum of T-invariant subspaces $\mathscr{M}_j$, as in Theorem 3.10:

$$\mathscr{V}_n = \mathscr{M}_1 \oplus \ldots \oplus \mathscr{M}_k, \qquad \text{where } \mathbf{T}(\mathscr{M}_j) \subseteq \mathscr{M}_j \text{ for each } j.$$

As in Exercise 6 of Section 3.4, if $M_j$ is a basis for $\mathscr{M}_j$ for $j = 1, \ldots, k$, then $M_1 \cup, \ldots, \cup M_k$ is a basis for $\mathscr{V}_n$. Since $\mathscr{M}_j$ is T-invariant, for each basis vector $\alpha$, if $\alpha \in M_j$ then $\mathbf{T}(\alpha) \in \mathscr{M}_j$. Hence in the matrix for $\mathbf{T}$ the column corresponding to $\mathbf{T}(\alpha)$ consists of zeros except in the rows corresponding to the basis vectors for $\mathscr{M}_j$. One of the most significant results in linear algebra is that any linear transformation can be represented by a diagonal block matrix in which each block is the sum of a scalar matrix and a nilpotent matrix having 0 or 1 in each position of the subdiagonal and zeros elsewhere.

*Nonsingular Matrix.*  An $n$-by-$n$ matrix $A$ is said to be *nonsingular* (or *invertible*) if and only if there exists a matrix $B$ such that $BA = I_n$. Otherwise $A$ is *singular.* Clearly $B$ must be $n$-by-$n$. Recall that a nonsingular mapping $\mathbf{T}$ from $\mathscr{V}_n$ to $\mathscr{W}$ is a one-to-one linear mapping of $\mathscr{V}_n$ onto $\mathscr{R}(\mathbf{T})$; hence we can represent $\mathbf{T}$ by an $n$-by-$n$ matrix $A$. From Theorem 3.8 and the isomorphism theorem, it follows that if $BA = I_n$, then $AB = I_n$. Hence

$B$ is called *inverse* of $A$ and is denoted $A^{-1}$. Later we shall derive several methods for calculating the inverse of a nonsingular matrix. Nonsingular matrices, of course, represent changes of bases, and therefore an $n$-by-$n$ matrix $A$ *is nonsingular if and only if its column vectors are linearly independent.*

*Orthogonal Matrix.* An orthogonal transformation $\mathbf{T}$ on Euclidean $n$-space preserves the inner product and thus describes a rigid motion of $E_n$. Relative to a normal orthogonal basis therefore, $\mathbf{T}$ must be represented by a matrix whose column vectors are of unit length and mutually orthogonal. Hence an $n$-by-$n$ matrix $A = (a_{ij})$ is said to be *orthogonal* if and only if for all $j, k = 1, \ldots, n$

$$\sum_{i=1}^{n} a_{ij} a_{ik} = \delta_{jk}.$$

Note that if we denote $a_{ij}$ by $b_{ji}$, then the matrix $B = (b_{ij})$ is such that $BA = I_n$. Hence an orthogonal matrix is nonsingular, and its inverse can be calculated very easily by writing the matrix $B$ whose rows are the columns of $A$. We shall return to this idea at the end of Chapter 5.

*Idempotent Matrix.* An $n$-by-$n$ matrix $A$ is said to be *idempotent* if and only if $A^2 = A$. Any projection is represented by an idempotent matrix. In fact, by choosing a basis as indicated in Theorem 3.12, a projection $\mathbf{T}$ can be represented by a diagonal matrix of the block form

$$\begin{pmatrix} I_{r(\mathbf{T})} & Z \\ Z & Z \end{pmatrix}.$$

Nondiagonal matrices can also be indempotent; for example,

$$\begin{pmatrix} 1 & 0 \\ -1 & 0 \end{pmatrix}.$$

*Nilpotent Matrix.* An $n$-by-$n$ matrix $A$ is said to be *nilpotent of index p* if and only if $A^p = Z$ but $A^{p-1} \neq Z$ for some integer $p > 1$. Any strictly triangular matrix is nilpotent. Furthermore, if $\mathbf{T}$ is a linear transformation on $\mathscr{V}_n$ and if $\mathbf{T}$ is nilpotent of index $n$, then relative to the basis described in Exercise 7 of Section 3.4, $\mathbf{T}$ is represented by the matrix

$$N = \begin{pmatrix} 0 & 0 & . & . & 0 & 0 \\ 1 & 0 & . & . & 0 & 0 \\ & . & . & & . & . \\ & . & . & & . & . \\ 0 & 0 & . & . & 0 & 0 \\ 0 & 0 & . & . & 1 & 0 \end{pmatrix},$$

which has 1 in each subdiagonal position and 0 elsewhere. $N$ is also nilpotent of index $n$.

*Unit Matrix.*  An $n$-by-$n$ matrix $U_{ij}$ is called a *unit matrix* if and only if 1 appears in the $(i, j)$ position and 0 appears in every other position. The effect of multiplying an $m$-by-$n$ matrix $A$ on the left by an $m$-by-$m$ unit matrix or on the right by an $n$-by-$n$ unit matrix may be calculated as an exercise.

## EXERCISES 3.6

1. Let $U_{ij}$ be a square matrix having 1 as the $(i, j)$ entry and 0 elsewhere, and let $A$ be any $m$-by-$n$ matrix.

(i)  Show that if $U_{ij}$ is $m$-by-$m$, then $U_{ij}A$ has zeros everywhere except in row $i$, and that row $i$ of $U_{ij}A$ is the same as row $j$ of $A$.

(ii)  Show that if $U_{ij}$ is $n$-by-$n$, then $AU_{ij}$ has zeros everywhere except in column $j$, and that column $j$ of $AU_{ij}$ is the same as column $i$ of $A$.

2. Extend the results of Exercise 1 to prove that if an $n$-by-$n$ matrix $A$ commutes with every $n$-by-$n$ matrix then $A$ is a scalar matrix.

3. Prove that an $n$-by-$n$ matrix $A$ commutes with every $n$-by-$n$ diagonal matrix if and only if $A$ is diagonal.

4. In the space $\mathcal{P}_3$ of polynomials of degree $\le 3$, choose $\{1, x, x^2, x^3\}$ as a basis, and let $\mathbf{D}$ be the derivative mapping: $\mathbf{D}p(x) = \dfrac{d}{dx} p(x)$. Represent $\mathbf{D}$ by a matrix $A$, and prove that $A$ is nilpotent. What is the index of nilpotency?

5. Describe precisely where the nonzero entries could occur in the product of two $n$-by-$n$ matrices, both of which are strictly lower triangular. Then make a reasonable conjecture about the product of any $k$ such matrices.

6. Prove that $A$ is nonsingular if and only if $\mathbf{T}_A$ is nonsingular, where $\mathbf{T}_A$ is

the linear transformation on $\mathcal{V}_n$ that corresponds to the $n$-by-$n$ matrix $A$ relative to a fixed but arbitrary basis.

7. Use Exercise 2 of Section 3.3 and the isomorphism theorem to show that if $A$ is a nonsingular $n$-by-$n$ matrix, then $A^{-1}$ is nonsingular and $(A^{-1})^{-1} = A$. Also if $c \neq 0$, then $(cA)^{-1} = c^{-1}A^{-1}$.

8. Prove that if $A$ and $B$ are nonsingular $n$-by-$n$ matrices, then $AB$ is nonsingular and $(AB)^{-1} = B^{-1}A^{-1}$.

9. For what values of $x$ is the following matrix singular?

$$\begin{pmatrix} 2 & -1 & 2 \\ -3 & 1 & -5 \\ 1 & 0 & x \end{pmatrix}$$

10. Do there exist values of $x$ and $y$ for which the following matrix is orthogonal? If so, find them.

$$\frac{1}{6}\begin{pmatrix} 1 & \sqrt{10} & x \\ -5 & \sqrt{10} & y \\ \sqrt{10} & 4 & \sqrt{10} \end{pmatrix}$$

11. Determine all real two-by-two idempotent matrices.

12. Determine all real two-by-two nilpotent matrices.

13. Determine all real two-by-two nonsingular matrices.

14. Determine all real two-by-two orthogonal matrices, and interpret each geometrically to determine the nature of all linear rigid motions of the Euclidean plane.

# 4

# Systems of Linear Equations

We now return to a general study of systems of $m$ linear equations in $n$ unknowns, considered in Section 1.4 for the special case in which $m = 3$. The methods introduced there are entirely applicable to the general case, so our main tasks in this chapter will be to formalize those methods, to develop the associated matrix theory, and to use our knowledge of linear mappings to give a simple geometric interpretation of the solutions of linear systems.

## 4.1 Gaussian Elimination

One of the most frequent applications of linear algebra arises from the need to solve systems of linear equations and systems of linear inequalities. Since any system of linear inequalities can be enlarged to a system of linear equations in which the newly introduced unknowns must be nonnegative, we shall study the problem of solving a system of $m$ linear equations in $n$ unknowns:

$$
\begin{aligned}
a_{11}x_1 + a_{12}x_2 + \ldots + a_{1n}x_n &= y_1, \\
a_{21}x_1 + a_{22}x_2 + \ldots + a_{2n}x_n &= y_2, \\
&\vdots \\
a_{m1}x_1 + a_{m2}x_2 + \ldots + a_{mn}x_n &= y_m.
\end{aligned}
$$

(4.1)

We consider the scalars $a_{ij}$ and $y_i$ as given, and we seek to determine *all* $n$-tuples $(x_1, \ldots, x_n)$ that satisfy the system. The system can be written more compactly in matrix notation as

$$AX = Y,$$

where

$$A = \begin{pmatrix} a_{11} & a_{12} & \cdots & a_{1n} \\ & & & \\ & \cdot & \cdot & \cdot \\ & & & \\ a_{m1} & a_{m2} & \cdots & a_{mn} \end{pmatrix}, \quad X = \begin{pmatrix} x_1 \\ \cdot \\ \cdot \\ \cdot \\ x_n \end{pmatrix}, \quad Y = \begin{pmatrix} y_1 \\ \cdot \\ \cdot \\ \cdot \\ y_m \end{pmatrix}.$$

We recall from Section 1.4 that the method of Gaussian elimination is to solve the first of these equations for $x_1$ in terms of $x_2, \ldots, x_n$ and to substitute that expression for $x_1$ into all of the other equations, thus exchanging the given system for a new system having the same set of solutions and in the form

$$x_1 + a_{11}^{-1}a_{12}x_2 + \ldots + a_{11}^{-1}a_{1n}x_n = a_{11}^{-1}y_1,$$
$$b_{22}x_2 + \ldots + b_{2n}x_n = u_2,$$
$$\vdots \qquad \qquad \vdots \qquad \vdots$$
$$b_{m2}x_2 + \ldots + b_{mn}x_n = u_m.$$

This exchange process is called a *pivot operation* with $a_{11}$ as *pivot*. A pivot, of course, must be nonzero; if $a_{11} = 0$ in the original system, we first rewrite the equations in a different order to guarantee that a nonzero coefficient occurs in the $(1, 1)$ position. The values of the new coefficients are given by

$$b_{ij} = a_{ij} - a_{11}^{-1}a_{i1}a_{1j}, \quad \text{for } i, j \geq 2,$$
$$u_i = y_i - a_{11}^{-1}a_{i1}u_1, \quad \text{for } i \geq 2.$$

Now the process is repeated, using the last $m - 1$ equations in $n - 1$ unknowns. If $b_{i2} = 0$ for all $i \geq 2$ then $x_2$ is missing from these $m - 1$ equations, and we move on to the next column to find the second pivot. The process stops when we run out of equations having nonzero coefficients.

As an example consider the three-by-four system

$$5x_2 + 2x_3 + 4x_4 = a,$$

$$x_1 + x_2 + x_3 + x_4 = b,$$

$$3x_1 - 2x_2 + x_3 - x_4 = c.$$

To represent the system by a matrix, we augment the matrix of coefficients by writing the constants $a$, $b$, and $c$ as an extra column on the right. Thus after interchanging the first and second equations we obtain the matrix

$$A_1 = \begin{pmatrix} 1^* & 1 & 1 & 1 & b \\ 0 & 5 & 2 & 4 & a \\ 3 & -2 & 1 & -1 & c \end{pmatrix}.$$

A pivot on the (1, 1) position produces the matrix

$$A_2 = \begin{pmatrix} 1 & 1 & 1 & 1 & b \\ 0 & 5^* & 2 & 4 & a \\ 0 & -5 & -2 & -4 & c - 3b \end{pmatrix}.$$

Next a pivot on the (1, 1) position of the indicated submatrix produces the matrix

$$A_3 = \begin{pmatrix} 1 & 1 & 1 & 1 & b \\ 0 & 1 & \dfrac{2}{5} & \dfrac{4}{5} & \dfrac{a}{5} \\ 0 & 0 & 0 & 0 & 5(c - 3b + a) \end{pmatrix}.$$

This last row corresponds to the equation

$$0x_1 + 0x_2 + 0x_3 + 0x_4 = 5(c - 3b + a).$$

Hence if $c - 3b + a \neq 0$, this system has no solution, and the original system likewise. To obtain a specific example of a consistent system, let us suppose that $a = 5$, $b = 1$, and $c = -2$. Then the last matrix becomes

$$A_3 = \begin{pmatrix} 1 & 1 & 1 & 1 & 1 \\ 0 & 1 & \dfrac{2}{5} & \dfrac{4}{5} & 1 \\ 0 & 0 & 0 & 0 & 0 \end{pmatrix}.$$

The solution can be computed by backward substitution

$$x_2 = 1 - \frac{2}{5}x_3 - \frac{4}{5}x_4,$$

$$x_1 = 1 - x_3 - x_4 - x_2$$

$$= 0 - \frac{3}{5}x_3 - \frac{1}{5}x_4,$$

or in vector form

$$\begin{pmatrix} x_1 \\ x_2 \end{pmatrix} = \begin{pmatrix} 0 \\ 1 \end{pmatrix} + x_3 \begin{pmatrix} -\frac{3}{5} \\ -\frac{2}{5} \end{pmatrix} + x_4 \begin{pmatrix} -\frac{1}{5} \\ -\frac{4}{5} \end{pmatrix}.$$

Any values can be chosen for $x_3$ and for $x_4$, and then corresponding values for $x_1$ and $x_2$ are determined.

A variation of this procedure, known as Gauss-Jordan elimination, performs the backward substitution as it goes along. In terms of the pivot operation this means that all pivots operate on the entire matrix instead of on only the remaining rows. In $A_2$ for example, if the pivot on the entry in the (2, 2) position were extended to the first row also, then the new matrix would be

$$B_3 = \begin{pmatrix} 1 & 0 & \frac{3}{5} & \frac{1}{5} & 0 \\ 0 & 1 & \frac{2}{5} & \frac{4}{5} & 1 \\ 0 & 0 & 0 & 0 & 0 \end{pmatrix},$$

from which the solution can be read directly. However, Gaussian elimination is somewhat more efficient than Gauss-Jordan elimination in terms of the amount of arithmetic involved.

In Section 1.4 it was shown that any pivot operation could be analyzed as a sequence of basic operations on the rows of a matrix:

(M)   Multiply a row by a nonzero constant.

(A)   Replace a given row by the sum of that row and another row.

If we adjoin a third operation,

(P)   Interchange any two rows,

then we see that both Gaussian and Gauss-Jordan elimination can be performed by finite sequences of these row operations.

Gaussian elimination performs such operations on a matrix until the resulting $m$-by-$n$ matrix is in *echelon form*:

(a) The first $k$ rows are nonzero and the last $m - k$ rows are zero, for some $k \leq m$;

(b) The first nonzero entry in each nonzero row is 1 and it occurs in a column to the right of the leading 1 in any preceding row.

The four-by-six matrix

$$M = \begin{pmatrix} 1 & 2 & 3 & 1 & 0 & 1 \\ 0 & 0 & 0 & 1 & 5 & 0 \\ 0 & 0 & 0 & 0 & 1 & -1 \\ 0 & 0 & 0 & 0 & 0 & 0 \end{pmatrix}$$

is in echelon form.

At this point Gaussian elimination stops and backward substitution is used to solve the system represented by the matrix.

It is clear, however, that further row operations can be used to produce 0 in each position above each of the leading 1's. In effect Gauss-Jordan elimination performs these additional row operations, bringing the final matrix into *reduced echelon form*. That is, in addition to being in echelon form, a matrix in reduced echelon form satisfies

(c) The first nonzero entry in each nonzero row is the only nonzero entry in its column.

The reduced echelon form of $M$ is, therefore,

$$R = \begin{pmatrix} 1 & 2 & 3 & 0 & 0 & -4 \\ 0 & 0 & 0 & 1 & 0 & 5 \\ 0 & 0 & 0 & 0 & 1 & -1 \\ 0 & 0 & 0 & 0 & 0 & 0 \end{pmatrix}.$$

It is an interesting fact that the *reduced* echelon form of a given matrix is uniquely determined by that matrix; that is, by performing row operations on a given matrix $A$ we can obtain one and only one matrix in reduced echelon form. To see heuristically why this result should be true, suppose that a matrix $A$ could be brought into reduced echelon form $R$ by one sequence of row operations and into a different reduced echelon matrix $R_1$ by another sequence of row operations. If we consider any system of linear equations for which $A$ is the coefficient matrix, then $R$ and $R_1$ are coefficient matrices of systems having precisely the same solutions. Since the solutions can be read directly from an augmented matrix in reduced echelon form, any difference in the coefficient matrices $R$ and $R_1$ will be reflected as a difference in the corresponding solutions.

## EXERCISES 4.1

1. Solve each of the following systems by means of Gaussian elimination.

(i)
$$\begin{aligned}
x_1 - x_2 + x_3 - x_4 + x_5 &= 1, \\
2x_1 - x_2 + 3x_3 \quad\quad + 4x_5 &= 2, \\
3x_1 - 2x_2 + 2x_3 + x_4 + x_5 &= 1, \\
x_1 \quad\quad + x_3 + 2x_4 + x_5 &= 0.
\end{aligned}$$

(ii)
$$\begin{aligned}
x_1 + 2x_2 + x_3 &= -1, \\
6x_1 + x_2 + x_3 &= -4, \\
2x_1 - 3x_2 - x_3 &= 0, \\
-x_1 - 7x_2 - 2x_3 &= 7, \\
x_1 - x_2 \quad\quad &= 1.
\end{aligned}$$

(iii)
$$\begin{aligned}
2x_1 + x_2 + 5x_3 &= 4, \\
3x_1 - 2x_2 + 2x_3 &= 2, \\
5x_1 - 8x_2 - 4x_3 &= 1.
\end{aligned}$$

(iv)
$$\begin{aligned}
x_1 + x_2 + x_3 + x_4 &= 1, \\
-x_1 \quad\quad + x_3 + 2x_4 &= 1, \\
3x_1 + 2x_2 \quad\quad - x_4 &= 1, \\
x_1 + x_2 + 2x_3 + 2x_4 &= 1.
\end{aligned}$$

(v)
$$\begin{aligned}
x_1 + 2x_2 - 3x_3 - 4x_4 &= 6, \\
x_1 + 3x_2 + x_3 - 2x_4 &= 4, \\
x_1 - x_2 - 5x_3 - 10x_4 &= 12.
\end{aligned}$$

2. Solve each of the systems in Exercise 1 by Gauss-Jordan elimination.

3. Use elementary row operations to determine the reduced echelon form of each of the following matrices.

(i)
$$\begin{pmatrix} 1 & 1 & 1 & 0 \\ -1 & 1 & 2 & 1 \\ 1 & 1 & 4 & 4 \end{pmatrix}.$$

(ii)
$$\begin{pmatrix} 0 & 3 & -2 & 1 & 0 \\ -3 & 0 & 1 & -4 & 1 \\ 2 & -1 & 0 & 0 & -2 \\ -1 & 4 & 0 & 0 & 1 \\ 0 & -1 & 2 & -1 & 0 \end{pmatrix}.$$

(iii)
$$\begin{pmatrix} 5 & 3 & 8 \\ 3 & 1 & 4 \\ -1 & 3 & 2 \end{pmatrix}.$$

(iv)
$$\begin{pmatrix} -1 & -3 & -2 \\ 5 & 7 & 2 \\ -3 & 1 & 4 \end{pmatrix}.$$

(v)
$$\begin{pmatrix} -1 & 5 & 6 \\ 1 & 2 & 1 \\ -1 & -3 & -2 \end{pmatrix}.$$

4.   In the linear system (4.1) select any nonzero coefficient, say $a_{rs}$. Gauss-Jordan elimination, using $a_{rs}$ as pivot, means that equation $r$ is solved for $x_s$ and the result substituted into all of the other equations to obtain an equivalent system in the form

$$BX = U.$$

Carry out the steps of this pivot operation to verify that

$$b_{ij} = \begin{cases} 1 & \text{if } i = r, j = s, \\ 0 & \text{if } i \neq r, j = s, \\ a_{rs}{}^{-1}a_{rj} & \text{if } i = r, j \neq s, \\ a_{rs}{}^{-1}\det \begin{pmatrix} a_{rs} & a_{rj} \\ a_{is} & a_{ij} \end{pmatrix} & \text{if } i \neq r, j \neq s, \end{cases}$$

where the two-by-two determinant is defined by

$$\det \begin{pmatrix} a & b \\ c & d \end{pmatrix} = ad - bc.$$

## 4.2   Elementary Row Operations

In the preceding section we have considered arithmetic procedures for solving a system of linear equations; in this section we shall develop these same ideas from the point of view of matrix algebra; and in Section 4.5 we shall reinterpret the problem geometrically by applying a knowledge of linear mappings. Our immediate objective is to formalize the concept of elementary row operations and to work out the corresponding matrix algebra.

**DEFINITION 4.1** The three *elementary row operations* on matrices are denoted as follows:

$M_i(c)$:  multiplication of row $i$ by $c \neq 0$,
$A_{ij}$:  addition of row $i$ to row $j$,
$P_{ij}$:  permutation of row $i$ and row $j$.

**DEFINITION 4.2** An *m-by-m elementary matrix* is any matrix that can be obtained by performing a single elementary row operation on the identity matrix $I$. The three types of $m$-by-$m$ elementary matrices will be denoted by $M_i(c)$, $A_{ij}$, and $P_{ij}$, according to the row operation used.

To describe the $m$-by-$m$ elementary matrices we recall from Section 3.6 that $U_{ij}$ denotes the unit $m$-by-$m$ matrix with 1 in the $(i, j)$ position and 0 elsewhere.

$$M_i(c) = \text{row } i \begin{pmatrix} 1 & & & & & & \\ & \ddots & & & & & \\ & & 1 & & & & \\ & & & c & & & \\ & & & & 1 & & \\ & & & & & \ddots & \\ & & & & & & 1 \end{pmatrix} = I + (c-1)U_{ii}.$$

$$A_{ij} = \begin{matrix} \\ \\ \text{row } i \\ \\ \\ \text{row } j \\ \\ \end{matrix} \begin{pmatrix} 1 & & & & & & \\ & \ddots & & & & & \\ & & 1 & & & & \\ & & & \ddots & & & \\ & & & & 1 & & \\ & & 1 & & & 1 & \\ & & & & & & \ddots \\ & & & & & & & 1 \end{pmatrix} = I + U_{ji}.$$

$$P_{ij} = \begin{array}{c} \\ \text{row } i \\ \\ \\ \\ \text{row } j \\ \\ \\ \end{array} \begin{pmatrix} 1 & & & & & & & & \\ & \ddots & & & & & & & \\ & & 1 & & & & & & \\ & & & 0 & & 1 & & & \\ & & & & 1 & & & & \\ & & & & & \ddots & & & \\ & & & & & & 1 & & \\ & & & 1 & & 0 & & & \\ & & & & & & & 1 & \\ & & & & & & & & \ddots \\ & & & & & & & & & 1 \end{pmatrix} = I - U_{ii} + U_{ij} - U_{jj} + U_{ji}.$$

The significance of elementary matrices is given by the following theorem.

**THEOREM 4.1**  Each elementary row operation can be performed on an $m$-by-$n$ matrix $A$ by multiplying $A$ on the *left* by the corresponding $m$-by-$m$ elementary matrix.
PROOF   Exercise.

This means that $P_{ij}A$ is the same as $A$ except that rows $i$ and $j$ have been interchanged; similarly for the other elementary row operations.

In order to perform elementary operations on the *columns* of an $m$-by-$n$ matrix $A$, only minor modifications are needed. Most important, $A$ must be multiplied on the *right* by an elementary matrix that must therefore be $n$-by-$n$. Furthermore, to add column $i$ to column $j$, we must multiply by $A_{ji}$ rather than $A_{ij}$. It is easy to verify directly that

$AM_i(c)$ is $A$ with column $i$ multiplied by $c$,
$AA_{ji}$ is $A$ with column $i$ added to column $j$,
$AP_{ij}$ is $A$ with columns $i$ and $j$ interchanged.

**THEOREM 4.2**  Each elementary matrix is nonsingular; $P_{ij}^{-1} = P_{ij}$, $M_i(c)^{-1} = M_i(c^{-1})$, and $A_{ij}^{-1} = M_i(-1)A_{ij}M_i(-1)$.
PROOF   The column vectors of each elementary matrix are linearly independent, so each such matrix is nonsingular. The form of each inverse is readily verified. In particular, we note that the inverse of an elementary matrix is a product of elementary matrices. This is simply the observation that each elementary row operation can be reversed by means of row operations.

**THEOREM 4.3**   For any $m$-by-$n$ matrix $A$, there exists a finite set of elementary matrices $E_1, \ldots , E_p$ such that

$$E_p \ldots E_1 A$$

is in reduced echelon form.

PROOF   Any computation algorithm for Gauss-Jordan elimination leads to a proof of this theorem. We already showed that $A$ can be converted to reduced echelon form by a sequence of elementary row operations and that each elementary row operation can be expressed as a product of elementary matrices operating to the left of $A$.

Now think of $A$ as representing a linear mapping **T**. By Definition 3.4 the rank of **T** is the dimension of the range space of **T**, which in turn is spanned by the column vectors of $A$. We define the *column rank* of $A$ to be the maximum number of linearly independent column vectors of $A$. Then clearly the column rank of $A$ is the number $r(\mathbf{T})$. If we interpret Theorem 3.9 by means of the isomorphism theorem as a theorem about matrices, we see that if $A$ is an $m$-by-$n$ matrix, if $B$ is a nonsingular $m$-by-$m$ matrix, and if $C$ is a nonsingular $n$-by-$n$ matrix, then the column ranks of the following matrices are equal: $A, BA, AC, BAC$.

From this and Theorem 4.3 it follows immediately that the column rank of $A$ is the same as the column rank of its reduced echelon form. However, for a matrix in reduced echelon form it is evident that every column is a linear combination of the unit columns; that is, those columns that contain the first nonzero entry of each nonzero row. Furthermore, these unit column vectors obviously are linearly independent, and the number of them is the number of nonzero rows. Also we observe that the nonzero rows form a maximal linearly independent set of row vectors of a matrix in reduced echelon form. You should be able to convince yourself of the general validity of these assertions by considering an example of a matrix in reduced echelon form:

$$\begin{pmatrix} 0 & 1 & * & 0 & 0 & * & * & 0 & * \\ 0 & 0 & 0 & 1 & 0 & * & * & 0 & * \\ 0 & 0 & 0 & 0 & 1 & * & * & 0 & * \\ 0 & 0 & 0 & 0 & 0 & 0 & 0 & 1 & * \\ 0 & 0 & 0 & 0 & 0 & 0 & 0 & 0 & 0 \end{pmatrix}.$$

Finally we note that each elementary row operation replaces one matrix by another having the same maximal number of linearly independent rows. This is obvious for operations $M_i(c)$ and $P_{ij}$. It follows for $A_{ij}$ if we regard the rows of $A$ as a set $\{\xi_1, \ldots, \xi_m\}$ of $m$ vectors in $R_n$. The row vectors of $A_{ij}A$ are $\{\xi_1, \ldots, \xi_{j-1}, \xi_j + \xi_i, \xi_{j+1}, \ldots, \xi_m\}$, and these two sets span the same subspace. To summarize:

In any $m$-by-$n$ matrix the maximal number of linearly independent row vectors equals the maximal number of linearly independent column vectors.

This discussion permits us to define the rank of a matrix and to draw a number of related conclusions.

**DEFINITION 4.3**  The *rank* $r(A)$ of an $m$-by-$n$ matrix $A$ is the maximal number of linearly independent columns of $A$.

**THEOREM 4.4**  Let $A$ be an $m$-by-$n$ matrix.

(a)  If $A$ represents a linear mapping **T** relative to chosen bases, then $r(\mathbf{T}) = r(A)$.

(b)  If $B$ is a nonsingular $m$-by-$m$ matrix and $C$ is a nonsingular $n$-by-$n$ matrix, then

$$r(A) = r(BA) = r(AC) = r(BAC).$$

(c)  The maximal number of linearly independent rows of $A$ is $r(A)$.

**THEOREM 4.5**  The following statements are equivalent for an $n$-by-$n$ matrix $A$:

(a)  $A$ is nonsingular.

(b)  $r(A) = n$.

(c)  The row vectors of $A$ are linearly independent.

(d)  The column vectors of $A$ are linearly independent.

# EXERCISES 4.2

1. Use elementary row operations to determine the rank of $A$.

(i)  $A = \begin{pmatrix} 3 & 1 & -2 & 4 \\ 2 & 0 & -5 & 1 \\ 1 & -1 & 2 & 6 \end{pmatrix}$ .

$$(ii)\ \ A = \begin{pmatrix} -2 & 1 & 4 & -2 & 3 \\ 1 & -5 & 2 & -3 & -2 \\ -4 & -7 & 16 & -12 & 5 \end{pmatrix}.$$

$$(iii)\ \ A = \begin{pmatrix} 1 & -1 & 2 & 1 \\ 4 & 3 & -1 & 0 \\ -2 & 2 & 1 & 7 \\ 2 & -9 & 3 & -10 \\ 9 & -2 & 4 & -4 \end{pmatrix}.$$

2. Show that the unit matrices satisfy $U_{hi}U_{jk} = \delta_{ij}U_{hk}$. Conclude that $U_{ij}$ is idempotent if $i = j$ and nilpotent of index 2 if $i \neq j$.

3. Prove Theorem 4.1.

4. Verify the three assertions made concerning the use of elementary matrices to perform elementary column operations on $A$.

5. Verify that the inverse of each elementary matrix is correctly stated in Theorem 4.2.

6. Write a sequence of elementary row operations whose only effect on $A$ is to add to row $j$ a constant multiple of row $i$.

7. Show that $P_{ij}$ can be expressed as a product of suitable matrices of the form $A_{rs}$ and $M_t(c)$.

## 4.3   Computation of $A^{-1}$

In the previous section we used the reduced echelon form of a matrix $A$ to conclude that the row rank of $A$ and the column rank of $A$ are always the same number. Another method is to study the transpose of a matrix, introduced for three-by-three matrices in our tentative look at quadratic forms in Section 1.12.

**DEFINITION 4.4**   If $A = (a_{ij})$ is an $m$-by-$n$ matrix, then the *transpose* $A^t$ of $A$ is the $n$-by-$m$ matrix defined by

$$A^t = (b_{ij}), \qquad \text{where } b_{ij} = a_{ji}.$$

Clearly the rows of $A$ are the columns of $A^t$, so if we could prove that the column rank of $A^t$ is the same as the column rank of $A$, we would have the desired result. This proof can be demonstrated by interpreting $A$ as a linear mapping $\mathbf{T}$ from $\mathcal{V}_n$ to $\mathcal{W}_m$, then showing that $A^t$ can be interpreted as an associated linear mapping $\mathbf{T}^t$ from the dual space of $\mathcal{W}_m$ to the dual space for $\mathcal{V}_n$, and finally proving that $r(\mathbf{T}) = r(\mathbf{T}^t)$. However, this approach would take us too far afield, and we shall not pursue it at this time. In any case, Theorem 4.4(c) provides a proof of the following result.

**THEOREM 4.6** For any $m$-by-$n$ matrix $A$,

$$r(A) = r(A^t).$$

**THEOREM 4.7** For any $m$-by-$n$ matrix $A$,

(a) $(cA)^t = cA^t$ for every scalar $c$,

(b) $(A + B)^t = A^t + B^t$ for every $m$-by-$n$ matrix $B$,

(c) $(A^t)^t = A$,

(d) $(AC)^t = C^tA^t$ for every $n$-by-$p$ matrix $C$.

PROOF The first three assertions are trivial consequences of Definition 4.4. To prove (d) we calculate the $(i, j)$ element of $(AC)^t$ as the $(j, i)$ element of $AC$, or

$$\sum_{k=1}^{n} a_{jk}c_{ki}.$$

But the $(i, k)$ element of $C^t$ is $c_{ki}$, and the $(k, j)$ element of $A^t$ is $a_{jk}$, so the $(i, j)$ element of $C^tA^t$ is

$$\sum_{k=1}^{n} c_{ki}a_{jk} = \sum_{k=1}^{n} a_{jk}c_{ki},$$

and so the transpose of a product is the product of the transposes *in the reverse order*.

As another application of the reduced echelon form, let us consider the special case of a nonsingular $n$-by-$n$ matrix $A$. Its rank is $n$, and its reduced echelon form must have $n$ columns that are unit vectors. This means that a sequence of elementary row operations reduces $A$ to the identity matrix:

(4.2) $$E_p \ldots E_1A = I.$$

Conversely, if any $n$-by-$n$ matrix $A$ can be reduced to $I$ by elementary row operations, so that (4.2) holds, then since the $E_i$ are nonsingular,

$$n = r(I) = r(E_p \, . \, . \, . \, E_1 A) = r(A),$$

so $A$ must be nonsingular. We include this result as part of the following extension of Theorem 4.5.

THEOREM 4.8    The following statements are equivalent for an $n$-by-$n$ matrix $A$:

(a)    $A$ is nonsingular.

(b)    The reduced echelon form of $A$ is $I$.

(c)    $A$ is the product of elementary matrices.

PROOF    We have already seen that (a) implies (b), which means that Equation (4.2) holds for suitable elementary matrices $E_1, \, . \, . \, . \, , E_p$. But since each elementary matrix is nonsingular,

$$A = E_1^{-1} E_2^{-1} \, . \, . \, . \, E_p^{-1}.$$

By Theorem 4.2 each $E_i^{-1}$ is a product of elementary matrices, and so is $A$. Hence (b) implies (c), and since the product of nonsingular matrices is nonsingular, (c) implies (a).

By multiplying (4.2) on each side by $A^{-1}$ we obtain

$$E_p \, . \, . \, . \, E_1 I = A^{-1}.$$

This equation provides us with our first method of actually computing the inverse of a nonsingular matrix:

Determine a sequence of elementary row operations that reduces $A$ to $I$. If the same sequence of row operations is applied to $I$, the resulting matrix is $A^{-1}$.

We can arrange the computations conveniently by writing $I$ alongside of $A$ to form an $n$-by-$2n$ matrix,

$$(I \mid A).$$

Perform row operations on this matrix to reduce $A$ to $I$, obtaining

$$(B \mid I).$$

Then $B = A^{-1}$. The following example illustrates the method.
To calculate the inverse of

$$A = \begin{pmatrix} 1 & 2 & 3 \\ 2 & 3 & 0 \\ 0 & 1 & 2 \end{pmatrix},$$

we write the block form $(I \mid A)$ and perform on this three-by-six matrix a sequence of row operations that reduces $A$ to $I$, yielding $(B \mid I)$. Then $B = A^{-1}$.

$$\begin{pmatrix} 1 & 0 & 0 & 1 & 2 & 3 \\ 0 & 1 & 0 & 2 & 3 & 0 \\ 0 & 0 & 1 & 0 & 1 & 2 \end{pmatrix} \xrightarrow{R_2-2R_1} \begin{pmatrix} 1 & 0 & 0 & 1 & 2 & 3 \\ -2 & 1 & 0 & 0 & -1 & -6 \\ 0 & 0 & 1 & 0 & 1 & 2 \end{pmatrix}$$

$$\xrightarrow{R_3+R_2} \begin{pmatrix} 1 & 0 & 0 & 1 & 2 & 3 \\ -2 & 1 & 0 & 0 & -1 & -6 \\ -2 & 1 & 1 & 0 & 0 & -4 \end{pmatrix} \xrightarrow[-\frac{1}{4}R_3]{-1R_2;} \begin{pmatrix} 1 & 0 & 0 & 1 & 2 & 3 \\ 2 & -1 & 0 & 0 & 1 & 6 \\ \frac{1}{2} & -\frac{1}{4} & -\frac{1}{4} & 0 & 0 & 1 \end{pmatrix}$$

$$\xrightarrow{R_2-6R_3} \begin{pmatrix} 1 & 0 & 0 & 1 & 2 & 3 \\ -1 & \frac{1}{2} & \frac{3}{2} & 0 & 1 & 0 \\ \frac{1}{2} & -\frac{1}{4} & -\frac{1}{4} & 0 & 0 & 1 \end{pmatrix}$$

$$\xrightarrow{R_1-2R_2-3R_3} \begin{pmatrix} \frac{3}{2} & -\frac{1}{4} & -\frac{9}{4} & 1 & 0 & 0 \\ -1 & \frac{1}{2} & \frac{3}{2} & 0 & 1 & 0 \\ \frac{1}{2} & -\frac{1}{4} & -\frac{1}{4} & 0 & 0 & 1 \end{pmatrix}.$$

Thus

$$A^{-1} = \frac{1}{4} \begin{pmatrix} 6 & -1 & -9 \\ -4 & 2 & 6 \\ 2 & -1 & -1 \end{pmatrix},$$

a result that should be checked by showing that $A^{-1}A = I$.
A modification of this method of computing $A^{-1}$ is developed in the next section, and a third method is described in Section 4.8.

## EXERCISES 4.3

1. Use elementary row operations to determine the rank of each of the following matrices and to calculate the inverse of any that is nonsingular.

(i)
$$A = \begin{pmatrix} -2 & 1 & 3 \\ 0 & -1 & 1 \\ 1 & 2 & 0 \end{pmatrix}.$$

(ii)
$$B = \begin{pmatrix} 1 & -1 & 1 & -1 \\ 0 & 1 & 0 & 1 \\ 1 & 0 & -1 & 0 \\ 0 & 1 & 0 & -1 \end{pmatrix}.$$

(iii)
$$C = \begin{pmatrix} 1 & 1 & 2 \\ 1 & 2 & 5 \\ 2 & 1 & 1 \end{pmatrix}.$$

(iv)
$$D = \begin{pmatrix} 4 & 2 & -1 \\ -5 & -3 & 1 \\ 3 & 2 & 0 \end{pmatrix}.$$

(v)
$$E = \begin{pmatrix} 1 & -2 & -3 \\ -2 & 0 & 4 \\ 1 & 1 & -1 \end{pmatrix}.$$

2. Calculate the transpose of each type of elementary matrix and observe that each is again an elementary matrix of the same type.

3. Use the results of Exercise 2 to verify the three statements about elementary column operations as listed immediately before Theorem 4.2.

4. Prove that if $A$ is a nonsingular square matrix, then $(A^t)^{-1} = (A^{-1})^t$.

## 4.4 Equivalence of Matrices

When an elementary row operation is applied to a system of linear equations, a new system is produced that has the same set of solutions as the original system. Systems having precisely the same solutions are said to be *equivalent* systems. We now define a similar concept for the matrices of equivalent systems.

**DEFINITION 4.5** An *m*-by-*n* matrix $B$ is said to be *row equivalent* to an *m*-by-*n* matrix $A$ if and only if $B$ can be derived by performing a finite sequence of elementary row operations on $A$.

Thus row equivalence is a binary relation, which we denote by $\sim$, on the set $\mathcal{M}_{m \times n}$ of all *m*-by-*n* matrices. Furthermore, row equivalence is

Reflexive:     $A \sim A$ for every $A$,
Symmetric:   If $B \sim A$, then $A \sim B$,
Transitive:   If $A \sim B$ and $B \sim C$, then $A \sim C$.

Reflexivity is obvious. Symmetry follows from the observation that if $B$ is obtained from $A$ by a finite sequence of elementary row operations,

$$B = E_p \ldots E_2 E_1 A,$$

where each $E_i$ denotes an elementary matrix, then

$$E_1^{-1} E_2^{-1} \ldots E_p^{-1} B = A.$$

However, by Theorem 4.2 the left hand side of this equation also describes a finite sequence of elementary row operations that, when applied to $B$, yields $A$. Transitivity follows from the equations

$$B = E_q \ldots E_2 E_1 C,$$

$$A = E_r \ldots E_{q+1} B.$$

Row equivalence, therefore, is a specific example of a general mathematical concept that is called an equivalence relation. Since a number of different equivalence relations are important in linear algebra, we digress briefly to investigate properties of equivalence relations defined on any nonvoid set.

An *equivalence relation* on a nonvoid set $S$ is a binary relation $\sim$ on $S$ which is reflexive, symmetric, and transitive.

Familiar examples of equivalence relations include equality of real numbers, similarity of triangles, congruence of integers modulo $n$, and parallelism of lines (if we agree that a line is parallel to itself.)

Given an equivalence relation $\sim$ on a set $S$, we can associate with each $x \in S$ the *equivalence class* $[x]$ of all elements of $S$ that are related to $x$ by $\sim$:

$$[x] = \{y \in S \,|\, y \sim x\}.$$

Clearly $x \in [x]$ because $\sim$ is reflexive, so each element of $S$ occurs in at least one equivalence class, and we now show that no element can appear in more than one equivalence class. Suppose that $y \in [x]$ and $y \in [z]$. Then $y \sim x$ and $y \sim z$. By symmetry and transitivity, $x \sim y$ and $y \sim z$, so $x \sim z$. Thus $x \in [z]$, and $z \in [x]$ by symmetry. By similar calculations which you should verify in detail, $[x] = [z]$. Hence the equivalence classes of $S$ are a collection of *disjoint* subsets whose union is $S$: *each element of S belongs to one and only one equivalence class.*

For any equivalence relation defined on the set $\mathcal{M}_{m \times n}$ of all $m$-by-$n$ matrices, we are interested in discovering a *canonical form,* a special (and preferably simple) type of matrix such that *each equivalence class contains one and only one matrix of that type.* This will enable us to determine whether any two matrices are in the same equivalence class by determining whether they have the same canonical form.

In the case of row equivalence, two systems of $m$ linear equations in $n$ variables have the same set of solutions if and only if the two $m$-by-$n$ matrices of coefficients are in the same row equivalence class. Furthermore, each row equivalence class of $m$-by-$n$ matrices contains exactly one matrix in reduced echelon form; that is, the reduced echelon form is canonical for row equivalence.

Row equivalence can also be characterized very simply by the following theorem.

**THEOREM 4.9**    Two $m$-by-$n$ matrices $A$ and $B$ are row equivalent if and only if there exists a nonsingular $m$-by-$m$ matrix $P$ such that

$$B = PA.$$

**PROOF**    If $A$ and $B$ are row equivalent, the nonsingular matrix $P$ is simply the product of the matrices that represent any sequence of elementary row operations used to transform $A$ into $B$. Conversely, if $B = PA$ the nonsingular matrix $P$ can be expressed as a product of elementary matrices by Theorem 4.8, and these elementary matrices define a sequence of elementary row operations that transforms $A$ into $B$.

If we consider a matrix as a rectangular array unrelated to the question of what such an array might represent, there is no reason to prefer row operations to column operations. Indeed we could develop a theory of column equivalence analagous to that of row equivalence. Instead of doing this, we shall now consider a theory that permits both elementary row operations and elementary column operations.

**DEFINITION 4.6**    An $m$-by-$n$ matrix $B$ is said to be *equivalent* to an $m$-by-$n$ matrix $A$ if and only if $B$ can be obtained by performing on $A$ a finite sequence of elementary row and column operations.

**THEOREM 4.10**   Equivalence of matrices is an equivalence relation; that is, it is reflexive, symmetric, and transitive.
PROOF   Exercise.

**THEOREM 4.11**   $B$ is equivalent to $A$ if and only if $B = PAQ$ for some nonsingular matrices $P$ and $Q$.
PROOF   From the remarks following Theorem 4.1 we know that elementary column operations are performed by multiplying on the right by suitable elementary matrices, and any product of such matrices in nonsingular. Theorem 4.7 then yields the assertion of this theorem.

We can now derive a simple canonical form for matrix equivalence.

**THEOREM 4.12**   An $m$-by-$n$ matrix $A$ of rank $r$ is equivalent to the $m$-by-$n$ matrix $B$ in which $b_{11} = b_{22} = \ldots = b_{rr} = 1$, and all other $b_{ij} = 0$. Two $m$-by-$n$ matrices are equivalent if and only if they have the same rank.
PROOF   $A$ is row equivalent to a reduced echelon matrix having $r$ distinct unit columns. Column permutations will place these unit columns in the first $r$ positions,

$$\left(\begin{array}{c|c} I_r & X \\ \hline Z & Z \end{array}\right).$$

Elementary column operations will then produce 0 in every entry of the upper right hand block to yield

$$B = PAQ = \left(\begin{array}{c|c} I_r & Z \\ \hline Z & Z \end{array}\right).$$

From Theorems 4.11 and 4.4(b), equivalent matrices have the same rank. Conversely if $A$ and $C$ are both of rank $r$, each is equivalent to the canonical matrix $B$, described above.

It follows that *a square matrix $A$ is nonsingular if and only if it is equivalent to $I$.* If $PAQ = I$, then $A = P^{-1}Q^{-1}$, and $A^{-1} = QP$. But $P$ describes the row operations and $Q$ the column operations used to reduce $A$ to $I$. This provides a second method for computing $A^{-1}$. We begin with the block form

$$\begin{array}{c|c} I & A \\ \hline & I \end{array},$$

and then apply row and column operations to this array to reduce the upper right hand block to $I$, obtaining

$$\begin{array}{c|c} P & I \\ \hline & Q \end{array}.$$

Then $QP = A^{-1}$.

## EXERCISES 4.4

1. Determine whether or not $A$ and $B$ are row equivalent, where

$$A = \begin{pmatrix} 3 & -1 & 2 \\ -2 & 1 & -1 \\ 3 & -2 & 1 \end{pmatrix}, \quad B = \begin{pmatrix} 2 & 1 & 1 \\ 1 & 2 & 1 \\ -1 & 4 & 1 \end{pmatrix}.$$

2. Which matrices in each of the following sets are equivalent?

   (i)  The matrices of Exercise 1.

   (ii)  The matrices of Exercise 1 of Section 4.3.

   (iii)  The matrices of Exercise 3 of Section 4.1.

3. Compute the inverse of each of the following matrices by the method of matrix equivalence.

   (i)  Exercise 1(i) of Section 4.3.

   (ii)  Exercise 1(iv) of Section 4.3.

   (iii)  Exercise 1(v) of Section 4.3.

4. Prove Theorem 4.10.

5. Relative to matrix equivalence, how many equivalence classes of $m$-by-$n$ matrices are there? Explain.

6. Given that $m$-by-$n$ matrices $A$ and $B$ are equivalent, determine whether each of the following pairs are equivalent:

   (i)  $A^t$ and $B^t$,

   (ii)  $A^2$ and $B^2$,

   (iii)  $AB$ and $BA$.

7. If $A$ is a real, symmetric $n$-by-$n$ matrix, show that $PAP^t$ is also real and symmetric for any real $m$-by-$n$ matrix $P$. (Symmetric means $A = A^t$.)

8. Let $A$ be a real, symmetric $n$-by-$n$ matrix. Carry out the steps indicated below to show that for some nonsingular matrix $P$, $PAP^t$ is a diagonal with $1, -1$, or $0$ in each diagonal position. Deduce that the number of nonzero entries of $PAP^t$ is $r(A)$.

(i)  Use Theorem 4.1 and the subsequent remarks to describe $EAE^t$ for each type of elementary matrix $E$; deduce that $EAE^t$ is symmetric and has the same rank as $A$.

(ii)  Use row and column operations to show that if $A \neq Z$, then $QAQ^t$ has a nonzero entry in the $(1, 1)$ position for some nonsingular $Q$.

(iii)  Use row and column operations to show that for some nonsingular $R$, $RAR^t$ has $0$ in each nondiagonal position of the first row and each nondiagonal position of the first column. Deduce also that if $A \neq Z$, then $R$ can be chosen so that the $(1, 1)$ entry is $1$ or $-1$.

(iv)  Complete the proof.

## 4.5  Geometry of Linear Systems

Having developed computational techniques for solving a system of $m$ linear equations in $n$ unknowns, we now examine such systems from the point of view of geometry. The system (4.1) can be represented in matrix form as

$$AX = Y,$$

where $A$ is an $m$-by-$n$ matrix, $X$ is an $n$-by-one matrix (column vector) and $Y$ is an $m$-by-one matrix (column vector). For a geometric interpretation, it is convenient to think of two Euclidean spaces $E_n$ and $E_m$, each equipped with a standard basis. Then $A$ represents a unique linear mapping $\mathbf{T}$ from $E_n$ to $E_m$, $X$ represents a vector $\xi$ in $E_n$, and $Y$ represents a vector $\eta$ in $E_m$. We regard the system (4.1) as specifying $\mathbf{T}$ and $\eta$; the problem is to determine all vectors $\xi \in E_n$ such that

$$\mathbf{T}\xi = \eta.$$

We distinguish two cases:

(a)  if $\eta = \theta$, the system is said to be *homogeneous*,
(b)  if $\eta \neq \theta$, the system is said to be *nonhomogeneous*.

The results that we shall derive for the existence of solutions, and the number and nature of the solutions if they exist, will depend upon three numbers: $n$, $r(\mathbf{T})$ (which equals $r(A)$), and the rank of the augmented matrix $(A|Y)$. The number $m$ is involved only through the relation $r(A) \le \min(m, n)$.

**THEOREM 4.13**  Let $AX = Y$ denote a system of $m$ linear equations in the $n$ variables $x_1, \ldots, x_n$.

(a)  If the system is *homogeneous* (that is, if $y_1 = \ldots = y_m = 0$), the solutions form a vector space of dimension $n - r(A)$. The zero solution $Z$ is called the *trivial* solution. Nontrivial solutions exist if and only if $r(A) < n$.

(b)  If the system is *nonhomogeneous* (that is, if some $y_j \ne 0$), a solution exists if and only if $r(A) = r(A|Y)$, where $(A|Y)$ is the augmented $m$-by-$(n+1)$ matrix obtained by adjoining the column $Y$ to the columns of $A$. If $X_0$ is a solution then $X$ is also a solution if and only if $X = X_0 + X_h$, where $X_h$ is a solution of the related homogeneous system $AX = Z$.

(c)  If $m = n = r(A)$, then for each $Y$ there is a unique solution of the system $AX = Y$.

**PROOF**  We use vector notation, $\mathbf{T}\xi = \eta$.

(a)  Let $\eta = \theta$. Then $\xi$ is a solution of the homogeneous system if and only if $\xi \in \mathscr{N}(\mathbf{T})$. But $\mathscr{N}(\mathbf{T})$ is a subspace of $\mathscr{V}_n$ of dimension $n(\mathbf{T}) = n - r(\mathbf{T}) = n - r(A) \ge 0$. The zero vector is always a solution; there exist nonzero solutions if and only if $r(A) < n$.

(b)  Let $\eta \ne \theta$. The solution of the system is the set of all $\xi$ that are mapped by $\mathbf{T}$ into $\eta$. Such $\xi$ will exist if and only if $\eta \in \mathscr{R}(\mathbf{T})$. But the space $\mathscr{R}(\mathbf{T})$ is spanned by the columns of $A$, so the system has a solution if and only if the column vector $Y$, corresponding to $\eta$, is a linear combination of the columns of $A$. This is equivalent to the statement $r(A|Y) = r(A)$.

Now suppose that this condition is satisfied and that $\xi_0$ is a fixed solution of the nonhomogeneous system. Then $\mathbf{T}(\xi_0) = \eta$. If $\xi$ is any solution, then $\mathbf{T}(\xi) = \eta$; so $\mathbf{T}(\xi - \xi_0) = \eta - \eta = \theta$, and $\xi - \xi_0 \in \mathscr{N}(\mathbf{T})$. Conversely, if $\xi - \xi_0 \in \mathscr{N}(\mathbf{T})$, $\xi$ is a solution. Hence any solution of the nonhomogeneous system is of the form $\xi_0 + \nu$, where $\nu \in \mathscr{N}(\mathbf{T})$. Since $\mathscr{N}(\mathbf{T})$ is a subspace of $\mathscr{V}_n$, the set $\{\xi_0 + \nu | \nu \in \mathscr{N}(\mathbf{T})\}$ of all solutions is a translation by the fixed vector $\xi_0$ of the subspace $\mathcal{N}(\mathbf{T})$, where $\mathcal{N}(\mathbf{T})$ is the set of all solutions of the corresponding homogeneous system.

(c)   If $m = n = r(A)$, then $A$ is a square matrix that is nonsingular. Hence $A^{-1}Y$ is a unique solution.

The relation $r(A|Y) = r(A)$ is called the *consistency condition* of the system; a system that satisfies the consistency condition is said to be *consistent*, and this is equivalent to saying that it has at least one solution. An *inconsistent* system is one for which no solution exists.

To illustrate for the case $n = 3$, the solution set of the linear system $AX = Y$ is void if the system is inconsistent. If the system is consistent, the solution set is a point or a line or a plane according as $r(A) = 3, 2$, or 1. Furthermore, the solution set contains the origin if and only if the system is homogeneous.

In summary, the solution set of a linear system $AX = Y$ is

(a)   void if and only if $r(A) < r(A|Y)$,

(b)   a translation of a subspace of $E_n$ of dimension $n - r(A)$ if and only if $r(A) = r(A|Y) \le n$, the translation vector being $\theta$ if and only if the system is homogeneous.

Note that Gaussian elimination (reduction of the augmented matrix to echelon form) actually exhibits these three numbers, leading either to the conclusion that the system is inconsistent or to a description of all solutions.

Further insight into the geometry can be obtained by considering a single linear equation in $n$ unknowns:

$$b_1 x_1 + b_2 x_2 + \ldots + b_n x_n = c.$$

In $E_n$ this represents the inner product $\langle \beta, \xi \rangle$ of two vectors. Given $\beta$, we ask for the set of all vectors $\xi$ whose inner product with $\beta$ equals the given number $c$. If $c = 0$ (homogeneous case), the solution vectors $\xi$ form the subspace of all vectors orthogonal to $\beta$, a subspace of dimension $n - 1$, which is called a *hyperplane* in $E_n$. If $c \neq 0$, the solution vectors $\xi$ are of the form $\xi_0 + \gamma$, where $\langle \beta, \xi_0 \rangle = c$ and $\langle \beta, \gamma \rangle = 0$. Geometrically, then, the solutions form a translated hyperplane. The solution of a system of $m$ linear equations, therefore, is the set of all points of intersection of $m$ translated hyperplanes. If there is no point common to all $m$ hyperplanes the system is inconsistent, and the converse is also true. If there is at least one point $\xi_0$ common to all $m$ translated hyperplanes, then the set of all solutions is the translation by $\xi_0$ of a subspace of dimension $n - r(A)$ (the intersection of the $m$ translated hyperplanes).

## EXERCISES 4.5

1. For each of the systems in Exercise 1 of Section 4.1 determine $m, n, r(A)$, and $r(A|Y)$ and reconcile these numbers with the solutions of that system.

2. Let $f$ be the function defined on $R_n$ by

$$f(x_1, \ldots, x_n) = b_1 x_1 + \ldots + b_n x_n,$$

where $b_1, b_2, \ldots, b_n$ are given real numbers, not all of which are zero.

   (i)   Verify that $f$ is a linear mapping from $R_n$ to $R$, or in the terminology of Section 3.1, a linear functional.

   (ii)   Determine the rank and nullity of $f$.

   (iii)   Deduce that the equation

$$b_1 x_1 + \ldots + b_n x_n = 0$$

defines a subspace of dimension $n - 1$ (a hyperplane) in $R_n$.

3. Show that if $m > n = r(A)$, then $AX = Y$ has either no solution or precisely one solution. Describe a computational process to distinguish these cases.

4. Determine a condition on $r(A)$ that is both necessary and sufficient that the system $AX = Y$ have a solution for all possible choices of $Y$. Prove your result.

5. Describe geometrically the possible nature of the solutions of the system

$$a_1 x_1 + a_2 x_2 + a_3 x_3 = d_1,$$
$$b_1 x_1 + b_2 x_2 + b_3 x_3 = d_2.$$

Relate your discussion to the consistency condition. What further modification is needed when a third equation,

$$c_1 x_1 + c_2 x_2 + c_3 x_3 = d_3,$$

is adjoined to the system?

6. What can you deduce about the number of solutions of a system of $m$ linear equations in $n$ unknowns if $m < n$? Take the consistency condition into account in your answer for the nonhomogeneous case.

## 4.6   Determinants

In the traditional school curriculum determinants are introduced in algebra courses as a method of solving systems of $n$ linear equations in $n$ unknowns, usually only for $n = 2$ and $n = 3$. For larger values of $n$,

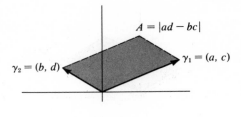

FIGURE 4.1

determinants are very inefficient, but they remain useful as a theoretical tool and a notational device in such subjects as linear algebra and multi-variable calculus.

Determinants also have an important geometric interpretation, as we noted briefly in Section 1.8, where in Euclidean three-space the triple scalar product $(\alpha \times \beta) \cdot \gamma$ was expressed as the determinant of the three row vectors that represent $\alpha$, $\beta$, and $\gamma$ in the standard coordinate system. Geometrically, the absolute value of $(\alpha \times \beta) \cdot \gamma$ is the *volume* of the parallelepiped having $\alpha$, $\beta$, and $\gamma$ as adjacent edges.

The determinant of a two-by-two matrix is defined algebraically by the formula

$$\det \begin{pmatrix} a & b \\ c & d \end{pmatrix} = ad - bc.$$

It is easy to verify that in $E_2$ the absolute value of this number is the *area* of the parallelogram having $\gamma_1 = a\epsilon_1 + c\epsilon_2$ and $\gamma_2 = b\epsilon_1 + d\epsilon_2$ as adjacent edges.

Although the case $n = 1$ is trivial, it is not without some interest. The determinant of a one-by-one matrix is defined in the obvious way,

$$\det (a) = a.$$

Again, we note that the absolute value of this number is the *length* in $E_1$ of the vector $\alpha = a\epsilon_1$. But length in $E_1$, area in $E_2$, and volume in $E_3$ all are measures of *geometric content* of a parallelotope having as adjacent edges as many vectors as the dimension of the space in which it is imbedded.

We wish to extract from these observations a few essential properties to guide us in defining the determinant in $E_n$. To begin with, each ordered set $C = \{\gamma_1, \ldots, \gamma_n\}$ of $n$ vectors in $E_n$ determines a scalar that is called the determinant of $C$. Relative to the standard coordinate system for $E_n$ each $\gamma_i$ is an $n$-tuple of scalars, and $C$ can be regarded as an $n$-by-$n$ matrix

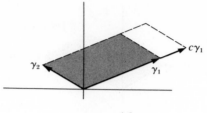

FIGURE 4.2

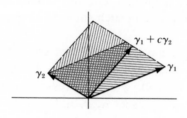

FIGURE 4.3

whose *columns* represent the vectors $\gamma_1, \ldots, \gamma_n$. Actually in our discussion of $n = 3$ in Section 1.8, we interpreted $C$ as the matrix whose *rows* are $\gamma_1, \gamma_2,$ and $\gamma_3$; but we can easily verify algebraically for $n = 1, 2, 3$ that det $C = $ det $C^t$, and the column interpretation is more convenient.

In $E_n$ the absolute value of det $C$ should provide the Euclidean measure of $n$-dimensional content of the $n$-parallelotope having $\gamma_1, \ldots, \gamma_n$ as adjacent edges. In particular, the following properties are reasonable expectations:

(1)  det $(\gamma_1, \ldots, \gamma_n) = 0$ if $\{\gamma_1, \ldots, \gamma_n\}$ is linearly dependent, since the parallelotope then collapses into a portion of a proper subspace of $n$-space;

(2)  if $p$ is any permutation of the indices $\{1, 2, \ldots, n\}$, then det $(\gamma_{p(1)}, \ldots, \gamma_{p(n)}) = \pm$ det $(\gamma_1, \ldots, \gamma_n)$, since both sets of vectors define the same $n$-parallelotope;

(3)  det $(c\gamma_1, \ldots, \gamma_n) = \pm c$det $(\gamma_1, \ldots, \gamma_n)$, since multiplying one edge of the $n$-parallelotope by $c$ should multiply its content by $|c|$ (see Figure 4.2);

(4)  det $(\gamma_1 + c\gamma_2, \gamma_2, \ldots, \gamma_n) = \pm$ det $(\gamma_1, \ldots, \gamma_n)$, since the two respective $n$-parallelotopes are constructed identically over two base parallelograms having the same area (see Figure 4.3);

(5)   det $(\epsilon_1, \ldots, \epsilon_n) = \pm 1$, since the $n$-dimensional content of a unit $n$-cube should be 1.

Properties (2), (3), and (4) are reminiscent of the three elementary row operations on matrices.

Rather surprisingly, it turns out that slight variations of Properties (3) and (4) are all that we need to derive a complete theory of $n$-by-$n$ determinants; Property (5) is needed only as a scalar factor to make the determinant unique, just as an agreement concerning the area of a one-by-one square is needed as a unit measure relative to which the area of more general figures can be expressed.

With this geometric motivation, we are now ready to state algebraically the definition of a determinant of an $n$-by-$n$ matrix $C$, which we shall regard as an $n$-tuple of column vectors $C_i$:

$$C = (C_1, \ldots, C_n).$$

**DEFINITION 4.7**  A *determinant* is a function (denoted det) that assigns to each $n$-by-$n$ matrix $C$ over $F$ a value det $C$ in $F$ such that the following three properties are satisfied: For each $c \in F$ and each $i = 1, \ldots, n$

(a)   det $(C_1, \ldots, cC_i, \ldots, C_n) = c$det $(C_1, \ldots, C_i, \ldots, C_n)$,

(b)   det $(C_1, \ldots, C_i + cC_j, \ldots, C_n) = $ det $(C_1, \ldots, C_i, \ldots, C_n)$ if $j \neq i$,

(c)   det $I = 1$.

Observe that there is no a priori guarantee that *at least one* such function exists for each $n$, nor is there any guarantee that *at most one* such function exists. It is easy to see that the customary notions of determinant for $n = 1, 2, 3$ satisfy this definition, so we shall proceed to investigate the properties of $n$-by-$n$ determinants on the assumption that at least one such function exists.

**THEOREM 4.14**  If det is a function having properties (a) and (b) of Definition 4.7, then

$$\text{det } (C_1, \ldots, C_i, \ldots, C_j, \ldots, C_n)$$
$$= -\text{det } (C_1, \ldots, C_j, \ldots, C_i, \ldots, C_n).$$

In words, det is an *alternating* function of the columns of $C$; or, an interchange of any two columns of $C$ reverses the sign of the determinant.

PROOF   Use Property (b) with $c = 1$, then with $c = -1$, and again with $c = 1$; then use (a):

$$
\begin{aligned}
\det C = &\ \det (C_1, \ldots, \quad C_i \ , \ldots, \qquad C_j \qquad , \ldots, C_n) \\
= &\ \det (C_1, \ldots, C_i + C_j, \ldots, \qquad C_j \qquad , \ldots, C_n) \\
= &\ \det (C_1, \ldots, C_i + C_j, \ldots, C_j - C_i - C_j, \ldots, C_n) \\
= &\ \det (C_1, \ldots, C_i + C_j, \ldots, \quad -C_i \qquad , \ldots, C_n) \\
= &\ \det (C_1, \ldots, \quad C_j \ , \ldots, \quad -C_i \qquad , \ldots, C_n) \\
= &\ -\det (C_1, \ldots, \quad C_j \ , \ldots, \qquad C_i \qquad , \ldots, C_n).
\end{aligned}
$$

Properties (1) and (2) as stated for determinants in $E_n$ follow immediately from Theorem 4.14; you may supply proofs as an exercise.

THEOREM 4.15   If the columns of $C$ are linearly dependent, then $\det C = 0$. If the columns of $B$ are a permutation of the columns of $C$, then $\det B = \pm \det C$, where the $+$ sign applies if the permutation can be performed by an even number of interchanges of pairs of columns, and the $-$ sign applies if the permutations can be performed by an odd number of interchanges of pairs of columns.
PROOF   Exercise.

THEOREM 4.16   If det is a function having Properties (a) and (b) of Definition 4.7, then for each $i = 1, \ldots, n$

$$\det (C_1, \ldots, A_i + B_i, \ldots, C_n)$$

$$= \det (C_1, \ldots, A_i, \ldots, C_n)$$

$$+ \det (C_1, \ldots, B_i, \ldots, C_n).$$

In words, if column $i$ of $C$ is expressed in any way as the sum of two column vectors, then det $C$ is the sum of two determinants as indicated.
PROOF   In view of Theorem 4.14, we need to prove only the case where $i = 1$. (Be sure that you understand why.) Hence, consider $\det (A_1 + B_1, C_2, \ldots, C_n)$. If the columns $C_2, \ldots, C_n$ form a linearly dependent set, then by Theorem 4.15 $\det (X, C_2, \ldots, C_n) = 0$ for any column $X$, and the conclusion of Theorem 4.16 holds. Otherwise,

we can choose a column vector $C_1$ so that the set $\{C_1, \ldots, C_n\}$ is a basis for the space of $n$-tuples over $F$. Then $A_1 = \sum_{i=1}^{n} a_i C_i$ and $B_1 = \sum_{i=1}^{n} b_i C_i$ for suitable scalars $a_i$ and $b_i$. Then from Properties (a) and (b),

$$\det (A_1 + B_1, C_2, \ldots, C_n)$$
$$= \det \left( \sum_{i=1}^{n} (a_i + b_i)C_i, C_2, \ldots, C_n \right)$$
$$= \det ((a_1 + b_1)C_1, C_2, \ldots, C_n)$$
$$= (a_1 + b_1) \det (C_1, C_2, \ldots, C_n)$$
$$= a_1 \det (C_1, C_2, \ldots, C_n) + b_1 \det (C_1, C_2, \ldots, C_n)$$
$$= \det (a_1 C_1, C_2, \ldots, C_n) + \det (b_1 C_1, C_2, \ldots, C_n)$$
$$= \det \left( \sum_{i=1}^{n} a_i C_i, C_2, \ldots, C_n \right) + \det \left( \sum_{i=1}^{n} b_i C_i, C_2, \ldots, C_n \right)$$
$$= \det (A_1, C_2, \ldots, C_n) + \det (B_1, C_2, \ldots, C_n).$$

By combining Theorem 4.16 with Property (a), we see that a determinant is a linear function of *each* of the columns of a matrix. That observation, the result of Theorem 4.14, and Property (c), frequently are chosen as the three defining properties of a determinant as a multilinear, alternating function $D$ from $\mathcal{V}_n \times \ldots \times \mathcal{V}_n$ ($n$ times) to $F$ such that $D(\epsilon_1, \ldots, \epsilon_n) = 1$. It is easy to see that the two definitions are equivalent; we preferred to begin with Definition 4.7 because of the geometric significance of those properties.

## EXERCISES 4.6

1. Show that in $E_2$ the absolute value of $\det A$ is the area of the parallelogram determined by the column vectors of $A$, relative to the $\epsilon$-basis.

2. Using the method given in Section 1.9, evaluate $\det A$, where

$$A = \begin{pmatrix} 3-x & 2 & -2 \\ 1 & 4-x & -1 \\ -2 & -4 & 1+x \end{pmatrix}.$$

For what values of $x$ is $A$ singular?

3. Show that a Cartesian equation for the line through two points $(a, b)$ and $(c, d)$ of the plane is

$$\det \begin{pmatrix} x & y & 1 \\ a & b & 1 \\ c & d & 1 \end{pmatrix} = 0.$$

Generalize to the case of a Cartesian equation for the plane through three points in space.

4. Prove Theorem 4.15.

5. Let $p$ be a permutation of $\{1, \ldots, n\}$. Suppose that the ordered set $(p(1), \ldots, p(n))$ can be returned to natural order by $k_1$ interchanges of pairs, and also in a different way by $k_2$ interchanges of pairs. Show that the existence of a determinant function implies that $k_1$ and $k_2$ are both odd or both even, depending on $p$.

## 4.7   Expansions for det $A$

Having derived the basic properties of determinants, we now apply those properties to derive specific formulas for the value of det $A$, still assuming that a determinant function exists.

Let $A$ and $B$ be $n$-by-$n$ matrices, and let $C = BA$. We see then that $\det C = \det (C_1, \ldots, C_n)$, where

$$C_k = \begin{pmatrix} c_{1k} \\ \cdot \\ \cdot \\ \cdot \\ c_{nk} \end{pmatrix}, \ c_{ik} = \sum_{j=1}^{n} b_{ij}a_{jk}, \ i = 1, \ldots, n.$$

That is,

$$c_{1k} = b_{11}a_{1k} + b_{12}a_{2k} + \ldots + b_{1n}a_{nk},$$
$$c_{2k} = b_{21}a_{1k} + b_{22}a_{2k} + \ldots + b_{2n}a_{nk},$$
$$\cdot \qquad\qquad\qquad\qquad \cdot$$
$$\cdot \qquad\qquad\qquad\qquad \cdot$$
$$\cdot \qquad\qquad\qquad\qquad \cdot$$
$$c_{nk} = b_{n1}a_{1k} + b_{n2}a_{2k} + \ldots + b_{nn}a_{nk},$$

or in column notation,

$$C_k = B_1 a_{1k} + B_2 a_{2k} + \ldots + B_n a_{nk} = \sum_{j=1}^{n} B_j a_{jk}.$$

Therefore we have

$$\det C = \det (C_1, \ldots, C_n)$$

$$= \det \left( \sum_{j_1=1}^{n} B_{j_1} a_{j_1 1}, \ldots, \sum_{j_n=1}^{n} B_{j_n} a_{j_n n} \right),$$

where each index $j_k$ runs independently from 1 to $n$. Since det is a linear function of each column, we can expand this last expression into the sum of $n^n$ determinants, most of which are zero:

$$\det C = \sum_{j_1=1}^{n} \ldots \sum_{j_n=1}^{n} (a_{j_1 1} \ldots a_{j_n n}) \det (B_{j_1}, \ldots, B_{j_n}).$$

Since $\det (B_{j_1}, \ldots, B_{j_n}) = 0$ whenever two subscripts coincide, nonzero terms occur only when the set of numbers $\{j_1, \ldots, j_n\}$ is a permutation of $\{1, \ldots, n\}$, say $\{p(1), \ldots, p(n)\}$;

$$\det C = \sum_{p} (a_{p(1)1} \ldots a_{p(n)n}) \det (B_{p(1)}, \ldots, B_{p(n)}),$$

where the summation extends over *all* permutations $p$ of $\{1, 2, \ldots, n\}$. The term $\det (B_{p(1)}, \ldots, B_{p(n)})$ is the determinant of a matrix whose columns are the columns of $B$, permuted somehow. By Theorem 4.15 its value is det $B$ if an *even* number of column transpositions restores the original order $(B_1, \ldots, B_n)$, but its value is $-$det $B$ if an *odd* number of column transpositions is required to restore the original order of columns. Hence

$$\det C = (\det B) \sum_{p} \pm a_{p(1)1} \ldots a_{p(n)n}.$$

Since this formula is valid for any $n$-by-$n$ matrices $A$ and $B$, it is valid for the particular case in which $B = I$. Then $C = A$, and by Definition 4.7 det $B = 1$, so we obtain the following theorem.

**THEOREM 4.17**   Any function det which has the three properties of a determinant specified in Definition 4.7 must have the algebraic form

$$\det A = \sum_{p} (-1)^{q} a_{p(1)1} a_{p(2)2} \ldots a_{p(n)n},$$

where the sum is extended over all permutations $p$ of $\{1, \ldots, n\}$ and $q$ is an integer which depends only on the permutation $p$.

The computation preceding the formal statement of Theorem 4.17 also yields a product formula for determinants.

**THEOREM 4.18**    det $(AB) = (\det A)(\det B)$.
PROOF    Exercise.

**THEOREM 4.19**    An $n$-by-$n$ matrix is nonsingular if and only if
$\det A \neq 0$. If $A$ is nonsingular, $\det (A^{-1}) = (\det A)^{-1}$.
PROOF    Exercise.

**THEOREM 4.20**    For every $n$-by-$n$ matrix $A$, $\det A^t = \det A$.
PROOF    Exercise. This theorem shows that all properties of determinants that are valid for columns are equally valid for rows.

Several points deserve special comment at this time. First, the role of Property (c) of Definition 4.7 as a scalar factor is now clear; until the last sentence before the statement of Theorem 4.17, our entire theory of determinants was derived from Properties (a) and (b). If we had specified arbitrarily that det $I$ should equal 27, then a new form of Theorem 4.17 would have been valid, with the factor 27 preceding the summation sign.

Second, *at most one* $n$-by-$n$ determinant function exists, since any such function must assign to $A$ the number specified by Theorem 4.17.

Third, although we now have a specific formula for det $A$, we do *not* propose to use it as a method of computation. Instead, when we must evaluate a determinant we will use the general properties developed in this section and an inductive expansion, which we now consider.

Let us fix a particular column, say column $j$, in the determinant of $A$. By Theorem 4.17 det $A$ can be expanded as a sum of $n!$ signed terms, each of which is a product of $n$ entries of $A$, *precisely one* from each row and each column. By collecting terms according to the elements of column $j$, we obtain an expression of the form

$$\det A = a_{1j}s_{1j} + \ldots + a_{ij}s_{ij} + \ldots + a_{nj}s_{nj},$$

where $s_{ij}$ denotes the sum of the $(n-1)!$ signed products of the form

$$\pm a_{p(1),1} \cdots a_{p(j-1),j-1}a_{p(j+1),j+1} \cdots a_{p(n),n}$$

such that $p(k) \neq i$ for all $k \neq j$. These are exactly the terms that occur (perhaps with different signs) in the term by term expansion of the determinant of the $(n-1)$-by-$(n-1)$ matrix $A_{ij}$ obtained from $A$ by deleting row $i$ and column $j$:

$$\det \begin{pmatrix} a_{11} & \cdots & a_{1j} & \cdots & a_{1n} \\ \cdot & & \cdot & & \cdot \\ \cdot & & \cdot & & \cdot \\ \cdot & & \cdot & & \cdot \\ a_{i1} & \cdots & a_{ij} & \cdots & a_{in} \\ \cdot & & \cdot & & \cdot \\ \cdot & & \cdot & & \cdot \\ \cdot & & \cdot & & \cdot \\ a_{n1} & \cdots & a_{nj} & \cdots & a_{nn} \end{pmatrix}$$

**DEFINITION 4.8**   For any $n$-by-$n$ matrix $A$ let $A_{ij}$ denote the $(n-1)$-by-$(n-1)$ matrix obtained by deleting row $i$ and column $j$ of $A$. The *cofactor* of $a_{ij}$ in $A$ is the scalar

$$\mathrm{cof}\ a_{ij} = (-1)^{i+j} \det A_{ij}.$$

If it so happens that this definition assigns the correct signs to the individual terms, then the sum $s_{ij}$ is simply the cofactor of $a_{ij}$ in $A$, and

(4.3)   $$\det A = \sum_{i=1}^{n} a_{ij}\ \mathrm{cof}\ a_{ij} \qquad \text{for each } j = 1, \ldots, n.$$

It is easy to check that this formula is correct for $n = 2$, since

$$\begin{aligned}
\det \begin{pmatrix} a_{11} & a_{12} \\ a_{21} & a_{22} \end{pmatrix} &= a_{11}a_{22} - a_{21}a_{12} \\
&= a_{11}\mathrm{cof}\ a_{11} + a_{21}\mathrm{cof}\ a_{21} \\
&= a_{12}\mathrm{cof}\ a_{12} + a_{22}\mathrm{cof}\ a_{22}.
\end{aligned}$$

Hence we are led to conjecture that (4.3) is valid for all $n > 0$.

Since a cofactor of an element in an $n$-by-$n$ matrix is an $(n-1)$-by-$(n-1)$ determinant, Formula (4.3) provides an inductive method to show that an $n$-by-$n$ determinant function exists for each $n$. We know that such a function exists for $n = 1$, so we make the inductive assumption that an $(n-1)$-by-$(n-1)$ determinant exists. Then we *define* a function $D$ on $\mathcal{M}_{n \times n}$ by

$$D(A) = \sum_{i=1}^{n} a_{ij}\ \mathrm{cof}\ a_{ij},$$

where $A$ is any $n$-by-$n$ matrix and $j$ is any fixed index from 1 to $n$. We can then verify that $D$ is a determinant function — that is, it satisfies Properties (a), (b), and (c) of Definition 4.7. Since we already know that such a function is unique if it exists, we can then conclude that (4.3) is indeed correct. The details of this proof are straight forward but rather long and mechanical, so we omit them.

**THEOREM 4.21**   An $n$-by-$n$ determinant function exists for every $n > 0$.

Formula (4.3) is known as the *Laplace expansion* of a determinant by means of the elements of column $j$. Since det $(A^t) = \det A$, the expansion can also be carried out by means of the elements of any row. In practice, one chooses a row or column having as many zeros as possible, thus reducing the number of cofactors to be evaluated. We now state a generalized form of the Laplace expansion.

**THEOREM 4.22**

$$\sum_{i=1}^{n} a_{ij} \operatorname{cof} a_{ik} = \delta_{jk} \det A,$$

$$\sum_{j=1}^{n} a_{ij} \operatorname{cof} a_{kj} = \delta_{ik} \det A.$$

PROOF    Exercise. For $j = k$ the first statement is simply (4.3), and the second statement follows from the first by Theorem 4.20.

Of course, the Kronecker delta reminds us of the identity matrix; if $A$ is nonsingular, the second formula of Theorem 4.22 shows that the $(i, k)$ entry of $I$ is

$$\sum_{j=1}^{n} a_{ij}(\det A)^{-1}(\operatorname{cof} a_{kj}),$$

which can be written as

$$\sum_{j=1}^{n} a_{ij}b_{jk} = \delta_{ik},$$

where

$$b_{jk} = (\det A)^{-1}\operatorname{cof} a_{kj}.$$

But $\sum_{j=1}^{n} a_{ij}b_{jk}$ is the $(i, k)$ entry of $AB$, so $AB = I$, and $B = A^{-1}$. We have therefore derived another method of calculating $A^{-1}$, as stated in the following theorem.

**THEOREM 4.23**   If $A$ is nonsingular, then the $(j,\,k)$ entry of $A^{-1}$ is

$$(\det A)^{-1}\mathrm{cof}\ a_{kj}.$$

That is, if $C$ is the matrix of cofactors of corresponding entries of $A$, then

$$A^{-1} = \frac{1}{\det A}\, C^{t}.$$

Note carefully that the matrix $C^{t}$ in Theorem 4.23 is the *transpose* of the matrix of cofactors of corresponding entries of $A$. We shall call the transposed matrix of cofactors of elements of $A$ the *comatrix* of $A$; frequently it is called the *adjoint* of $A$, but the term adjoint is also used in another context in linear algebra, so the term comatrix is used to avoid ambiguity. Thus

$$\mathrm{com}\ A = (b_{ij}), \qquad \text{where } b_{ij} = \mathrm{cof}\ a_{ji},$$

$$A^{-1} = \frac{1}{\det A}\ \mathrm{com}\ A.$$

Therefore, Theorem 4.23 provides another method for computing the inverse of a nonsingular matrix. Although it is a good method for $n = 2$ or $n = 3$, the method of elementary row operations is more efficient if $n > 3$.

EXERCISES 4.7

1. Prove that the determinant of a lower triangular matrix is the product of the diagonal elements.

2. Prove Theorem 4.18.

3. Prove Theorem 4.19.

4. Calculate the determinant of each of the three types of $n$-by-$n$ elementary matrices, and show that $\det E^{t} = \det E$ for every elementary matrix $E$.

5. Use Exercise 4 to prove Theorem 4.20.

6. Illustrate Theorem 4.22 by calculating $\sum_{j=1}^{3} a_{ij}\,\mathrm{cof}\ a_{kj}$ for $i = 1$ and $k = 2$, and for $i = 2$, $k = 2$, given that

$$A = \begin{pmatrix} 1 & -1 & 2 \\ -2 & 3 & 1 \\ 2 & -2 & x \end{pmatrix}.$$

7. Prove Theorem 4.22.

8. Evaluate by using properties of determinants to produce a row or column having 0 as most of its entries and then expanding by the elements of that row or column:

$$\det \begin{pmatrix} 3 & 1 & -2 & 4 \\ 2 & 0 & -5 & 1 \\ 1 & -1 & 2 & 6 \\ -2 & 3 & -2 & 3 \end{pmatrix}.$$

9. Let $A$ be an $n$-by-$n$ matrix which can be partitioned into blocks,

$$A = \begin{pmatrix} B & Z \\ C & D \end{pmatrix}$$

in such a way that $B$ is $m$-by-$m$ and $Z$ consists entirely of zeros. Deduce that $\det A = (\det B)(\det D)$.

## 4.8   Some Applications of Determinants (*An Optional Section*)

In this section we shall examine briefly three applications of determinants in linear algebra and calculus: Cramer's rule for the solution of a system of $n$ linear equations in $n$ variables, the Wronskian determinant in the solution of linear differential equations, and the Jacobian determinant in multiple integration.

First consider a system of $n$ linear equations in $n$ variables,

$$AX = Y,$$

where we assume that the $n$-by-$n$ matrix $A$ is nonsingular. Then the column vectors $\{A_1, \ldots, A_n\}$ are linearly independent, and $Y$ is a linear combination of these columns. By writing the equations in extended form,

$$y_1 = a_{11}x_1 + a_{12}x_2 + \ldots + a_{1n}x_n,$$
$$y_2 = a_{21}x_1 + a_{22}x_2 + \ldots + a_{2n}x_n,$$
$$\vdots \qquad\qquad\qquad\qquad \vdots$$
$$y_n = a_{n1}x_1 + a_{n2}x_2 + \ldots + a_{nn}x_n,$$

we see that $Y = x_1 A_1 + x_2 A_2 + \ldots + x_n A_n$. Thus the coefficients of this linear combination are the components of the unique solution vector $X$.

Now for any fixed $j$, consider the determinant of the matrix obtained by writing $Y$ in place of column $j$ in $A$:

$$\det (A_1, \ldots, A_{j-1}, Y, A_{j+1}, \ldots, A_n)$$

$$= \det (A_1, \ldots, \sum_{i=1}^{n} x_i A_i, \ldots, A_n)$$

$$= \sum_{i=1}^{n} x_i \det (A_1, \ldots, A_i, \ldots, A_n)$$

$$= x_j \det A,$$

since each determinant in the sum is 0 except where $i = j$. The conclusion

$$x_j = \frac{\det (A_1, \ldots, Y, \ldots, A_n)}{\det (A_1, \ldots, A_j, \ldots, A_n)}$$

is *Cramer's rule* for solving an $n$-by-$n$ linear system by means of determinants. Computationally it is grossly inefficient for $n > 2$ and should not be used.

Determinants commonly arise in the study of calculus in two situations. In the study of integration of functions of several variables, a suitable change of variables defined by $x_i = g_i(u_1, \ldots, u_n), i = 1, \ldots, n$, carries a region $U$ of $E_n$ into a region $X$ of $E_n$, and transforms an integral over $X$ into an integral over $U$ according to the formula

$$\int\!\!\int \ldots \int_{X} f(x_1, \ldots, x_n) dx_1 dx_2 \ldots dx_n$$

$$= \int\!\!\int \ldots \int_{U} f(g_1, \ldots, g_n) |J| du_1 du_2 \ldots du_n.$$

The expression $|J|$ is the absolute value of an $n$-by-$n$ determinant of partial derivatives, called the Jacobian determinant:

$$J = \det \begin{pmatrix} \dfrac{\partial g_1}{\partial u_1} & \cdots & \dfrac{\partial g_1}{\partial u_n} \\[1em] \cdot & & \cdot \\ \cdot & & \cdot \\ \cdot & & \cdot \\[1em] \dfrac{\partial g_n}{\partial u_1} & \cdots & \dfrac{\partial g_n}{\partial u_n} \end{pmatrix},$$

This integration formula is the $n$-dimensional integral form of the chain rule. Geometrically $J$ adjusts for the distortion of volume ($n$-dimensional content) introduced by the mapping (perhaps nonlinear) that defines the change of variables.

Another application concerns solutions of linear homogeneous differential equations:

$$(4.4) \qquad y^{(n)} + a_1(x)y^{(n-1)} + \ldots + a_{n-1}(x)y' + a_n(x)y = 0,$$

where $a_1(x), \ldots, a_n(x)$ are continuous over some interval $I$ of the real line. The set of all solutions is a vector space, and it can be shown that the dimension of the solution space is $n$. Hence a problem of major importance is to determine a basis for the solution space, and therefore to know when a set of solutions is linearly independent. A set of nonzero real functions $\{y_1, \ldots, y_k\}$ is linearly dependent over an interval $I$ if and only if there exist scalars $c_i$ not all zero such that

$$\sum_{i=1}^{k} c_i y_i$$

is the *zero function* over $I$. Assuming these functions to be sufficiently differentiable on $I$, we have for all $x \in I$

$$(4.5) \qquad \begin{aligned} c_1 y_1(x) & + \ldots + c_k y_k(x) & = 0, \\ c_1 y_1'(x) & + \ldots + c_k y_k'(x) & = 0, \\ & \vdots \\ c_1 y_1^{(k-1)}(x) & + \ldots + c_k y_k^{(k-1)}(x) & = 0. \end{aligned}$$

The determinant

$$W(x) = \det \begin{pmatrix} y_1 & \cdot & \cdot & \cdot & y_k \\ y_1' & \cdot & \cdot & \cdot & y_k' \\ \cdot & & & & \cdot \\ \cdot & & & & \cdot \\ y_1^{(k-1)} & \cdot & \cdot & \cdot & y_k^{(k-1)} \end{pmatrix}$$

is called the *Wronskian* of $\{y_1, \ldots, y_k\}$. Suppose that for some $x_0 \in I$, $W(x_0) \neq 0$. Then the system of linear equations in $c_1, \ldots, c_k$ obtained when $x = x_0$ in (4.5) has only the trivial solution: $c_1 = c_2 = \ldots = c_k = 0$. Hence if $\{y_1, \ldots, y_k\}$ is linearly dependent over $I$, the Wronskian must vanish identically over $I$.

If we also assume that each of these functions is a solution of the differential equation (4.4), where $k \le n$, the converse can be proved: if the Wronskian vanishes identically over $I$, then $\{y_1, \ldots, y_k\}$ is linearly dependent. However, as an exercise you may show that if the functions are not solutions of an equation of the form (4.4), the Wronskian can vanish identically without the functions being linearly dependent.

## EXERCISES 4.8

1. Solve the following system of linear equations,
$$x + 2y + 8z = 1,$$
$$-x + y + 5z = -3,$$
$$2x + 2y + 4z = 5,$$

   (i)  by Cramer's rule,

   (ii)  by the comatrix method of calculating $A^{-1}$,

   (iii)  by reduction to reduced echelon form.

2. Do the same as in Exercise 1 for the system
$$2x - y + 3z = 3,$$
$$y = -2,$$
$$2x + y + z = 1.$$

3. (i)  Show that if $A$ is $n$-by-$n$, then $\det(cA) = c^n \det A$.

   (ii)  Show that every skew symmetric matrix of odd dimension is singular. (Skew symmetric means $A^t = -A$.)

4. Prove that $\det(\text{com } A) = (\det A)^{n-1}$, where $A$ is $n$-by-$n$ and com $A$ is the comatrix of $A$. Deduce that $A$ is singular if and only if com $A$ is singular.

5. The Vandermonde matrix of order $n$ is, by definition,

$$V(x_1, \ldots, x_n) = \begin{pmatrix} 1 & 1 & \ldots & 1 \\ x_1 & x_2 & \ldots & x_n \\ x_1^2 & x_2^2 & \ldots & x_n^2 \\ \cdot & \cdot & & \cdot \\ \cdot & \cdot & & \cdot \\ \cdot & \cdot & & \cdot \\ x_1^{n-1} & x_2^{n-1} & \ldots & x_n^{n-1} \end{pmatrix}.$$

   (i)  For $n = 2, 3$ verify that $\det V = \Pi_{1 \le i < j \le n}(x_j - x_i)$, where $\Pi$ denotes "product."

   (ii)  Prove the statement in (i) for all $n > 1$.

6. Use Exercise 5 to show that $A$ is nonsingular, where

$$
A = \begin{pmatrix}
1^0 & 1^1 & \cdots & 1^{n-1} \\
2^0 & 2^1 & \cdots & 2^{n-1} \\
\cdot & \cdot & & \cdot \\
\cdot & \cdot & & \cdot \\
\cdot & \cdot & & \cdot \\
(n-1)^0 & (n-1)^1 & \cdots & (n-1)^{n-1} \\
n^0 & n^1 & \cdots & n^{n-1}
\end{pmatrix}.
$$

7. Calculate the Wronskian of each of the following sets of functions. Which sets are linearly independent?

(i) $\{e^{ax}, e^{bx}\}, a \neq b$.

(ii) $\{\sin bx, \cos bx\}, b \neq 0$.

(iii) $\{2, \sin^2 x, \cos 2x\}$.

(iv) $\{\sinh; x, \cosh; x, e^{-x}\}$.

8. Let $y_1 = x^2$ and $y_2 = x|x|$. Show that $y_1$ and $y_2$ are differentiable on $[-1, 1]$, the Wronskian of $y_1$ and $y_2$ vanishes identically on $[-1, 1]$, and yet $\{y_1, y_2\}$ is linearly independent.

9. The familiar method of integration by substitution asserts that if $x = g(y)$, $a = g(c)$, $b = g(d)$, and if $g$ is monotone on $[c, d]$, then

$$
\int_a^b f(x)dx = \int_c^d f(g(y))g'(y)dy.
$$

Is this formula consistent with the multivariable formula stated in this section? Explain.

10. An integration formula for changing from rectangular to polar coordinates is

$$
\iint_X f(x, y)dxdy = \iint_U f(r\cos\theta, r\sin\theta) \, r \, drd\theta.
$$

Demonstrate in detail how this formula is derived from the analogous formula stated in this section.

11. Apply the substitution $x = u^2 - v^2$, $y = 2uv$ to express

$$
\iint_X \sqrt{x^2 + y^2} \, dxdy
$$

as a double integral over a portion of the $u$-$v$ plane.

# 5

# *Diagonalization*

*(An optional chapter)*

In this chapter we resume the study of linear transformations and quadratic forms, giving particular attention to problems of diagonal representation, introduced tentatively in Chapter 1. Out of pedagogical necessity these topics are often recognized as optional in a first course in linear algebra in spite of their importance in more advanced developments of the subject.

## 5.1   Similarity of Matrices

Given a fixed linear transformation $\mathbf{T}$ from $\mathscr{V}_n$ to $\mathscr{V}_n$, we shall be interested in determining intrinsic geometric properties of $\mathbf{T}$ and in relating these properties to the algebraic properties of any matrix that represents $\mathbf{T}$. One of the major problems is to choose a basis for $\mathscr{V}_n$ so that the matrix representing $\mathbf{T}$ is as simple as possible. Analogous considerations pertain to quadratic forms.

First, however, we reconsider matrix equivalence to see how this relation, which arose naturally in the matrix representation of systems of linear equations, also has significance for linear mappings. An $m$-by-$n$ matrix $C$ is equivalent to an $m$-by-$n$ matrix $A$ if and only if there exist

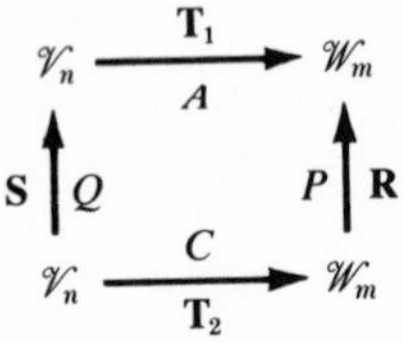

FIGURE 5.1

nonsingular matrices $M$ and $Q$ such that

$$C = MAQ,$$

or in equivalent form,

$$M^{-1}C = AQ.$$

Letting $P = M^{-1}$, we can state the condition for the equivalence of $A$ and $C$ to be the existence of a nonsingular $m$-by-$m$ matrix $P$ and a non-singular $n$-by-$n$ matrix $Q$ such that

$$PC = AQ.$$

$A$ and $C$ are $m$-by-$n$ and, hence, represent linear mappings, say $\mathbf{T}_1$ and $\mathbf{T}_2$, from $\mathscr{V}_n$ to $\mathscr{W}_m$. $P$ represents a nonsingular linear transformation $\mathbf{R}$ (a change of basis) in $\mathscr{W}_m$, and $Q$ represents $\mathbf{S}$, a change of basis in $\mathscr{V}_n$. This situation is shown schematically in Figure 5.1, and the relation between $\mathbf{T}_1$ and $\mathbf{T}_2$ is established in Theorem 5.1.

THEOREM 5.1  Two $m$-by-$n$ matrices $A$ and $C$ are equivalent if and only if they represent the same linear mapping from $\mathscr{V}_n$ to $\mathscr{W}_m$ relative to two suitably chosen pairs of bases.

PROOF  Assume that $A$ and $C$ are equivalent and that $PC = AQ$. Choose any basis $\{\alpha_1, \ldots, \alpha_n\}$ for $\mathscr{V}_n$ and any basis $\{\beta_1, \ldots, \beta_m\}$ for $\mathscr{W}_m$, and let $\mathbf{T}_1$ be the mapping represented by $A$ relative to that pair of bases:

$$\mathbf{T}_1(\alpha_k) = \sum_{i=1}^{m} a_{ik}\beta_i.$$

For $j = 1, \ldots, n$ let $\gamma_j \in \mathscr{V}_n$ be defined by column $j$ of $Q$:

$$\gamma_j = \sum_{k=1}^{n} q_{kj}\alpha_k.$$

Since $Q$ is nonsingular, $\{\gamma_1, \ldots, \gamma_n\}$ is a basis for $\mathscr{V}_n$. Thus $Q$ is the matrix that expresses each vector of the $\gamma$-basis in terms of the vectors of the $\alpha$-basis. The identity mapping $\mathbf{S}$ in $\mathscr{V}_n$,

$$\mathbf{S}(\gamma_j) = \gamma_j = \sum_{k=1}^{n} q_{kj}\alpha_k,$$

is represented by $Q$ relative to the *pair* $(\gamma, \alpha)$ of bases for $\mathscr{V}_n$. Similarly, in $\mathscr{W}_m$ for $k = 1, \ldots, m$, let $\delta_k$ be defined by column $k$ of $P$,

$$\delta_k = \sum_{i=1}^{m} p_{ik}\beta_i.$$

Since $P$ is nonsingular, $\{\delta_1, \ldots, \delta_m\}$ is a basis for $\mathscr{W}_m$, and relative to the *pair* $(\delta, \beta)$ of bases for $\mathscr{W}_m$ the identity mapping $\mathbf{R}$ is represented by $P$:

$$\mathbf{R}(\delta_k) = \delta_k = \sum_{i=1}^{m} p_{ik}\beta_i.$$

Finally, let $\mathbf{T}_2$ be the mapping from $\mathscr{V}_n$ to $\mathscr{W}_m$ that is represented by $C$ relative to the $\gamma$-basis for $\mathscr{V}_n$ and the $\delta$-basis for $\mathscr{W}_m$:

$$\mathbf{T}_2(\gamma_j) = \sum_{k=1}^{m} c_{kj}\delta_k = \sum_{k=1}^{m} c_{kj}\left(\sum_{i=1}^{m} p_{ik}\beta_i\right)$$

$$= \sum_{i=1}^{m}\left(\sum_{k=1}^{m} p_{ik}c_{kj}\right)\beta_i.$$

Since $PC = AQ$, the coefficient of $\beta_i$ in this last expression is the $(i, j)$ entry of both $PC$ and $AQ$. Hence

$$\mathbf{T}_2(\gamma_j) = \sum_{i=1}^{m}\left(\sum_{k=1}^{n} a_{ik}q_{kj}\right)\beta_i$$

$$= \sum_{k=1}^{n} q_{kj}\left(\sum_{i=1}^{m} a_{ik}\beta_i\right) = \sum_{k=1}^{n} q_{kj}\mathbf{T}_1(\alpha_k)$$

$$= \mathbf{T}_1\left(\sum_{k=1}^{n} q_{kj}\alpha_k\right) = \mathbf{T}_1(\gamma_j).$$

Hence $\mathbf{T}_1 = \mathbf{T}_2$, as claimed. Conversely, if $A$ and $C$ both represent the same linear mapping $\mathbf{T}$ relative to two pairs of bases $(\alpha, \beta)$ and $(\gamma, \delta)$ for $\mathscr{V}_n$ and $\mathscr{W}_m$, let $\mathbf{R}$ and $\mathbf{S}$ be defined as above; $\mathbf{R}$ expresses each $\delta$ as a linear combination of the $\beta_i$, and $\mathbf{S}$ expresses each $\gamma$ as a linear combination of the $\alpha_k$, thus defining the matrices $P$ and $Q$. The previous calculations, carried out in a different order, then show that $PC = AQ$.

A similar analysis also provides an answer to the following analogous question: if a given matrix $A$ represents the linear mapping $T_1$ relative to a pair $(\alpha, \beta)$ of bases for $\mathscr{V}_n$ and $\mathscr{W}_m$ and the same matrix represents $T_2$ relative to another pair $(\gamma, \delta)$ of bases, how are $T_1$ and $T_2$ related? In this case, if nonsingular linear transformations $\mathbf{Q}$ and $\mathbf{P}$ are defined by

$$\mathbf{Q}(\gamma_k) = \alpha_k, \qquad \text{for } k = 1, \ldots, n,$$

$$\mathbf{P}(\delta_i) = \beta_i, \qquad \text{for } i = 1, \ldots, m,$$

an easy calculation shows that $\mathbf{PT_2} = \mathbf{T_1Q}$.

For the rest of this chapter our attention will be restricted to linear transformations on $\mathscr{V}_n$ and to $n$-by-$n$ matrices. Then in Theorem 5.1, $\mathscr{W}_m = \mathscr{V}_n$, and we can specify that $\alpha_i = \beta_i$ and $\gamma_i = \delta_i$ for $i = 1, \ldots, n$. Then $\mathbf{R} = \mathbf{S}$ and $P = Q$, so the matrices $A$ and $C$ are related by the equation $PC = AP$, or $C = P^{-1}AP$.

**DEFINITION 5.1**   An $n$-by-$n$ matrix $C$ is said to be *similar* to an $n$-by-$n$ matrix $A$ if and only if there exists a nonsingular $n$-by-$n$ matrix $P$ such that

$$C = P^{-1}AP.$$

**THEOREM 5.2**   Similarity is an equivalence relation on $\mathscr{M}_{n \times n}$.
**PROOF**   Exercise.

**THEOREM 5.3**   If $A$ and $C$ are similar then

$$r(A) = r(C),$$

$$\det A = \det C.$$

**PROOF**   Exercise. The converse is false.

**THEOREM 5.4**   Two $n$-by-$n$ matrices $A$ and $C$ are similar if and only if they represent the same linear transformation on $\mathscr{V}_n$, each relative to a suitably chosen basis for $\mathscr{V}_n$.
**PROOF**   This statement is a special case of Theorem 5.1 in which $\alpha_i = \beta_i$ and $\gamma_i = \delta_i$ for $i = 1, \ldots, n$, $\mathbf{R} = \mathbf{S}$, and $P = Q$.

It is important to realize that similarity is obtained by specializing equivalence in two ways: in requiring that $m = n$ and also in the selection of bases. Although both equivalence and similarity are defined for $n$-by-$n$ matrices, they are different relations and define different equivalence classes. Similar matrices are equivalent, but equivalent $n$-by-$n$ matrices need not be similar. From Theorem 5.4 we see that each linear transformation **T** on $\mathcal{V}_n$ determines a similarity class of matrices, the set of all matrices that represent **T** relative to the various possible choices of a basis. In order to represent **T** as simply as possible, we wish to select from that similarity class a matrix having a simple form. In short, we seek a canonical form for similarity—a special form of $n$-by-$n$ matrix such that *one and only one* such matrix belongs to each similarity class. Since a full derivation of a canonical form for similarity lies beyond the normal expectations and time limits of a first course in linear algebra, it is not included here. However, in the next two sections we shall explore this problem and obtain some important results related to it.

For emphasis we now summarize our results concerning the matrices that can represent a given linear transformation **T** on $\mathcal{V}_n$. If $\{\alpha_1, \ldots, \alpha_n\}$ is a basis for $\mathcal{V}_n$, **T** is represented relative to that basis by the matrix $A$ in which, for $j = 1, \ldots, n$, column $j$ is the $n$-tuple of scalars that expresses $T\alpha_j$ as a linear combination of $\{\alpha_1, \ldots, \alpha_n\}$. Relative to a different basis $\{\beta_1, \ldots, \beta_n\}$ for $\mathcal{V}_n$, **T** is represented by a matrix $B$ that can be computed from $A$ as follows: let $P$ be the $n$-by-$n$ matrix in which, for $j = 1, \ldots, n$, column $j$ is the $n$-tuple of scalars that expresses $\beta_j$ as a linear combination of $\{\alpha_1, \ldots, \alpha_n\}$. Then

$$B = P^{-1}AP.$$

## EXERCISES 5.1

1. Prove Theorem 5.2.

2. Prove Theorem 5.3.

3. Apply Theorems 5.3 and 4.12 to write two two-by-two matrices that are equivalent but not similar.

4. Let **T** be the linear mapping of $\mathcal{V}_3$ into $\mathcal{W}_2$ whose matrix relative to a pair $(\alpha, \beta)$ of bases is

$$A = \begin{pmatrix} 1 & 0 & -3 \\ 2 & 1 & 1 \end{pmatrix}.$$

Let new bases be defined for $\mathscr{V}_3$ and $\mathscr{W}_2$ respectively by

$$\gamma_1 = \alpha_1 + \alpha_2 + \alpha_3,$$
$$\gamma_2 = \qquad \alpha_2 + \alpha_3,$$
$$\gamma_3 = \alpha_1 \qquad + \alpha_3,$$

and

$$\delta_1 = \beta_1 - 2\beta_2,$$
$$\delta_2 = \beta_1 + \beta_2.$$

Compute directly the matrix $C$ that represents $\mathbf{T}$ relative to the pair $(\gamma, \delta)$ of bases. Also compute the matrices $P$ and $Q$ of Theorem 5.1 and verify that $C = P^{-1}AQ$.

5. In the Euclidean plane, let $\epsilon_1 = (1, 0)$, $\epsilon_2 = (0, 1)$, $\alpha_1 = (1, 1)$, and $\alpha_2 = (1, -1)$. If a linear transformation $\mathbf{T}$ is represented relative to the $\epsilon$-basis by the matrix

$$A = \begin{pmatrix} 1 & 1 \\ 1 & \cdot 1 \end{pmatrix},$$

find the matrix $B$ that represents $\mathbf{T}$ relative to the $\alpha$-basis, and find $P$ such that $B = P^{-1}AP$. Check your result by matrix computations.

6. Let $\mathbf{T}$ be the linear transformation of $\mathscr{V}_3$ that is represented relative to a basis $\{\alpha_1, \alpha_2, \alpha_3\}$ by

$$A = \begin{pmatrix} 1 & 2 & 1 \\ 2 & 0 & -2 \\ -1 & 2 & 3 \end{pmatrix}.$$

(i)  Show that another basis for $\mathscr{V}_3$ is defined by

$$\gamma_1 = \alpha_1 + \alpha_2,$$
$$\gamma_2 = \alpha_1 \qquad + \alpha_3,$$
$$\gamma_3 = \alpha_1 - \alpha_2 + \alpha_3.$$

(ii)  Determine the matrix $C$ that represents $\mathbf{T}$ relative to the $\gamma$-basis by computing $\mathbf{T}(\gamma_j)$ in terms of the $\gamma$'s.

(iii)  Write a matrix $Q$ that represents the change of coordinates $\mathbf{S}$ from the $\gamma$-basis to the $\alpha$-basis

$$\mathbf{S}\gamma_j = \gamma_j = \sum_{k=1}^{3} q_{kj}\alpha_k,$$

and verify that $C = Q^{-1}AQ$.

7. Recalling the discussion in Section 3.6, show that any two $n$-by-$n$ idempotent matrices of the same rank are similar. Describe a form for idempotent matrices that is canonical with respect to similarity.

8. Are $AB$ and $BA$ similar for all $n$-by-$n$ matrices $A$ and $B$? What can be said if $A$ is nonsingular?

9. If $A$ and $B$ are similar, determine which of the following pairs are similar:

(i)  $cA$ and $cB$ for any scalar $c$,

(ii)  $A^k$ and $B^k$, $k = 1, 2, \ldots$ ,

(iii)  $A^{-1}$ and $B^{-1}$, if $A$ is nonsingular,

(iv)  $A^t$ and $B^t$.

10. Use Exercise 9 to show that if $A$ and $B$ are similar, so are $p(A)$ and $p(B)$, where $p$ is any polynomial.

11. In the analysis of three-phase power systems an impedance matrix occurs in the form

$$C = \begin{pmatrix} z_1 & z_3 & z_2 \\ z_2 & z_1 & z_3 \\ z_3 & z_2 & z_1 \end{pmatrix},$$

where each $z_j$ is a complex number. Let $e = (1/2)(-1 + i\sqrt{3})$, where $i^2 = -1$. (Observe that $e^3 = 1$ and $e^2 + e + 1 = 0$.) Show by computation that $P^{-1}CP$ is diagonal, where

$$P = \begin{pmatrix} 1 & 1 & 1 \\ 1 & e & e^2 \\ 1 & e^2 & e \end{pmatrix}.$$

12. Let $A$ represent a linear transformation $\mathbf{T}_1$ relative to an $\alpha$-basis for $\mathscr{V}_n$ and a linear transformation $\mathbf{T}_2$ relative to a $\gamma$-basis. Define a nonsingular linear transformation $\mathbf{S}$ on $\mathscr{V}_n$ such that

$$\mathbf{T}_2 = \mathbf{S}^{-1}\mathbf{T}_1\mathbf{S},$$

and verify this equation by detailed calculations.

## 5.2   Characteristic Vectors

Each similarity class of $n$-by-$n$ matrices consists of precisely those matrices that represent the same linear transformation $\mathbf{T}$ on $\mathscr{V}_n$, relative to various bases for $\mathscr{V}_n$. All of these matrices, therefore, must share a number of properties reflecting the geometric properties that characterize $\mathbf{T}$. To look for a simple matrix representation of $\mathbf{T}$, therefore, is to look for a basis for $\mathscr{V}_n$ that is intrinsically related to $\mathbf{T}$.

We begin by looking for vectors that are mapped by $\mathbf{T}$ in the simplest possible way. Of course $\mathbf{T}(v) = \theta$ for each $v \in \mathscr{N}(\mathbf{T})$; but if $\mathbf{T}$ is nonsingular, $\mathcal{N}(\mathbf{T}) = [\theta]$, so this is no help in general. We might also look for a fixed point—a vector $\alpha$ such that $\mathbf{T}(\alpha) = \alpha$. But again this is too special. More generally we look for a nonzero vector that is mapped into a scalar multiple of itself:

$$\mathbf{T}(\alpha) = c\alpha \text{ for some } \alpha \neq \theta, \text{ and some scalar } c.$$

If $c = 0$, $\alpha \in \mathcal{N}(\mathbf{T})$; if $c = 1$, $\alpha$ is a fixed point of $\mathbf{T}$.

**DEFINITION 5.2**    A nonzero vector $\alpha$ is called a *characteristic vector* of $\mathbf{T}$ if and only if there exists a scalar $c$ such that

$$\mathbf{T}(\alpha) = c\alpha.$$

The associated scalar $c$ is called a *characteristic value* of $\mathbf{T}$.

Other common terms are eigenvector and eigenvalue, proper vector and proper value. The set of all characteristic values of $\mathbf{T}$ is called the *spectrum* of $\mathbf{T}$. According to widely accepted notation, the Greek letter $\lambda$ is used to denote a characteristic value of $\mathbf{T}$, and we shall use that notation, even though it is an exception to our convention of using Greek letters for vectors and Latin letters for scalars.

Relative to a basis for $\mathcal{V}_n$, $\mathbf{T}$ is represented by an $n$-by-$n$ matrix $A$ and $\xi$ by a column vector $X$. Then $\xi$ is a characteristic vector of $\mathbf{T}$, associated with the characteristic value $\lambda$, if and only if

$$\mathbf{T}\xi = \lambda\xi,$$

$$AX = \lambda X,$$

$$(A - \lambda I)X = Z,$$

where $Z$ is the zero column vector. This last equation is the matrix form of a homogeneous system of $n$ linear equations in the $n$ unknowns $x_i$, and by Theorem 4.13(a) a nonzero solution vector $X$ exists if and only if $r(A - \lambda I) < n$. This is equivalent to the condition

$$\det (A - \lambda I) = 0.$$

From our knowledge of determinants we see that

$$\det \begin{pmatrix} a_{11}-\lambda & a_{12} & \ldots & a_{1n} \\ a_{21} & a_{22}-\lambda & \ldots & a_{2n} \\ \cdot & \cdot & & \cdot \\ \cdot & \cdot & & \cdot \\ \cdot & \cdot & & \cdot \\ a_{n1} & a_{n2} & \ldots & a_{nn}-\lambda \end{pmatrix}$$

is a polynomial of degree $n$ in the scalar variable $\lambda$:

$$p(\lambda) = \det(A - \lambda I) = (-1)^n(\lambda^n + b_1\lambda^{n-1} + \ldots + b_n),$$

where each $b_i$ is an algebraic sum of products of entries of $A$. The characteristic values of **T** are those values of $\lambda$ for which $p(\lambda) = 0$. Since we need to be assured that all the zeros of any such polynomial are elements of the scalar field, we shall assume that *the scalar field is the field of complex numbers*. Then, at least in theory, we can factor $p(\lambda)$ as a product of linear factors:

$$p(\lambda) = (-1)^n(\lambda - \lambda_1) \ldots (\lambda - \lambda_n),$$

where the $\lambda_i$ are complex numbers, *not necessarily distinct*. Collecting like linear factors, we have

$$p(\lambda) = (-1)^n(\lambda - \lambda_1)^{s_1} \ldots (\lambda - \lambda_t)^{s_t},$$

where $s_1 + \ldots + s_t = n$, and $\lambda_1, \ldots, \lambda_t$ are the distinct characteristic values.

**DEFINITION 5.3** The polynomial $\det(A - \lambda I)$ is called the *characteristic polynomial* of the matrix $A$. The equation $\det(A - \lambda I) = 0$ is called the *characteristic equation* of $A$. Any solution $\lambda_0$ of the characteristic equation of $A$ is called a *characteristic value* of $A$, and any column vector $X_0$ for which $AX_0 = \lambda_0 X_0$ is called a *characteristic vector* of $A$, associated with the characteristic value $\lambda_0$.

Our discussion has established the following result:

**THEOREM 5.5** The characteristic values of a linear transformation **T** are the characteristic values of any matrix that represents **T**.

**THEOREM 5.6**    If $A$ and $B$ are similar, they have the same characteristic polynomial and hence the same set of characteristic values.
PROOF    If $B = P^{-1}AP$, then

$$B - \lambda I = P^{-1}AP - \lambda I = P^{-1}(A - \lambda I)P.$$

Hence $B - \lambda I$ and $A - \lambda I$ are similar and therefore have the same determinant.

The converse of Theorem 5.6 is not true; matrices having the same characteristic polynomial are not necessarily similar. This fact makes a canonical form for similarity somewhat more complicated than we might wish.

As an example let $\mathbf{T}$ be the linear transformation on $E_3$ that is represented relative to the standard $\epsilon$-basis by

$$A = \begin{pmatrix} 1 & 2 & 1 \\ 2 & 0 & -2 \\ -1 & 2 & 3 \end{pmatrix}.$$

The characteristic polynomial is

$$p(\lambda) = \det \begin{pmatrix} 1-\lambda & 2 & 1 \\ 2 & -\lambda & -2 \\ -1 & 2 & 3-\lambda \end{pmatrix} = -\lambda(\lambda - 2)^2.$$

The distinct characteristic values are 0 and 2. $X$ is a column vector such that $(A - \lambda I)X = Z$ if and only if

$$(1 - \lambda)x_1 + 2x_2 + \qquad\quad x_3 = 0,$$
$$2x_1 - \lambda x_2 - \qquad\quad 2x_3 = 0,$$
$$-x_1 + 2x_2 + (3 - \lambda)x_3 = 0.$$

For $\lambda = 0$, the solutions of this system are vectors of the form

$$X = \begin{pmatrix} a \\ -a \\ a \end{pmatrix} \qquad \text{for any } a.$$

For $\lambda = 2$, the solutions are of the form

$$X = \begin{pmatrix} b \\ 0 \\ b \end{pmatrix} \qquad \text{for any } b.$$

To be specific we choose $a = 1 = b$ and obtain two linearly independent characteristic vectors, which we label as shown:

$$X_3 = \begin{pmatrix} 1 \\ -1 \\ 1 \end{pmatrix} \qquad \text{and} \qquad X_2 = \begin{pmatrix} 1 \\ 0 \\ 1 \end{pmatrix}.$$

Any characteristic vector of $A$ is a scalar multiple of $X_3$ or $X_2$, and any nonzero multiple of either is a characteristic vector. We observe that a basis consisting entirely of characteristic vectors simply does not exist. However, if we choose

$$X_1 = \begin{pmatrix} 1 \\ 1 \\ 0 \end{pmatrix},$$

then $\{X_1, X_2, X_3\}$ is a basis such that

$$AX_1 = \begin{pmatrix} 3 \\ 2 \\ 1 \end{pmatrix} = 2X_1 + X_2,$$

$$AX_2 = 2X_2,$$

$$AX_3 = 0X_3.$$

Hence, relative to this basis, **T** is represented by the matrix

$$C = \begin{pmatrix} 2 & 0 & 0 \\ 1 & 2 & 0 \\ 0 & 0 & 0 \end{pmatrix}.$$

An alternative method for obtaining $C$ was described in the final paragraph of Section 5.1. For this example we have

$$P = \begin{pmatrix} 1 & 1 & 1 \\ 1 & 0 & -1 \\ 0 & 1 & 1 \end{pmatrix},$$

from which we can compute

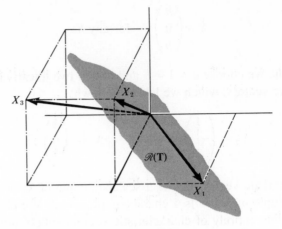

FIGURE 5.2

$$P^{-1} = \begin{pmatrix} 1 & 0 & -1 \\ -1 & 1 & 2 \\ 1 & -1 & -1 \end{pmatrix},$$

and then verify that $C = P^{-1}AP$. It turns out that $C$ provides the simplest possible representation of **T**.

The preceding example suggests that the set of all characteristic vectors associated with a single characteristic value is a subspace of $\mathscr{V}_n$—in this example, each such subspace is one-dimensional, but that is not the case in general. We also observe in the preceding example that any two characteristic vectors associated with *distinct* characteristic values are linearly independent.

**THEOREM 5.7** Let **T** be a linear transformation on $\mathscr{V}_n$ and let $\lambda_0$ be a fixed characteristic value of **T**. The set $\mathscr{C}_0$ of all characteristic vectors associated with $\lambda_0$, together with the zero vector, forms a **T**-invariant subspace of $\mathscr{V}_n$.

**PROOF** Exercise. $\mathscr{C}_0$ is called the *characteristic subspace* associated with $\lambda_0$.

**THEOREM 5.8** If $\lambda_1$ and $\lambda_2$ are distinct characteristic values of **T**, then $\mathscr{C}_1 \cap \mathscr{C}_2 = [\theta]$.

**PROOF** If $\eta \in \mathscr{C}_1 \cap \mathscr{C}_2$, then $\mathbf{T}\eta = \lambda_1\eta = \lambda_2\eta$, so $(\lambda_1 - \lambda_2)\,\eta = \theta$. Since $\lambda_1 - \lambda_2 \neq 0$, $\eta = \theta$.

This very simple theorem has very important consequences, since it shows that the sum of two characteristic subspaces associated with distinct characteristic values is a *direct* sum. (See Definition 2.4.) By Exercise 10 of Section 2.3, if $B_1$ is a basis for $\mathscr{C}_1$ and $B_2$ a basis for $\mathscr{C}_2$, then $B_1 \cup B_2$ is a basis for $\mathscr{C}_1 \oplus \mathscr{C}_2$. Moreover, Theorem 5.8 can be extended by an inductive argument to show that if $\lambda_1, \ldots, \lambda_k$ are distinct characteristic values, then

$$\mathscr{C}_1 \cap (\mathscr{C}_2 + \ldots + \mathscr{C}_k) = [\theta].$$

This result implies that the sum of distinct characteristic subspaces is a direct sum:

$$\mathscr{C}_1 + \ldots + \mathscr{C}_k = \mathscr{C}_1 \oplus \ldots \oplus \mathscr{C}_k.$$

In particular, if $\xi_i \in \mathscr{C}_i$, then $\{\xi_1, \ldots, \xi_k\}$ is linearly independent; and if $B_i$ is a basis for $\mathscr{C}_i$, then $B_1 \cup \ldots \cup B_k$ is a basis for $\mathscr{C}_1 \oplus \ldots \oplus \mathscr{C}_k$. We shall apply these remarks in the next section.

EXERCISES 5.2

1. Find the characteristic polynomial, characteristic values, and a maximal linearly independent set of characteristic vectors of each of the following matrices:

(i)
$$\begin{pmatrix} 0 & 2 \\ 3 & -1 \end{pmatrix},$$

(ii)
$$\begin{pmatrix} 3 & 2 & 4 \\ 2 & 0 & 2 \\ 4 & 2 & 3 \end{pmatrix},$$

(iii)
$$\begin{pmatrix} 3 & 2 & 1 & 0 \\ 0 & 1 & 0 & 1 \\ 0 & 2 & 1 & 0 \\ 0 & 0 & 0 & 1 \end{pmatrix}.$$

2. Given the matrix

$$A = \begin{pmatrix} 2 & 1 & 1 \\ -2 & 1 & 3 \\ 3 & 1 & -1 \end{pmatrix}.$$

(i) Find the characteristic polynomial and three distinct characteristic values of $A$.

(ii) Show that $\alpha_1 = (5, 1, 4)$, $\alpha_2 = (3, -5, 2)$, and $\alpha_3 = (0, 1, -1)$ are a basis of characteristic vectors.

(iii) Calculate $P^{-1}AP$, where the column vectors of $P$ are $\alpha_1, \alpha_2, \alpha_3$.

(iv) Reread Example (d) of Section 1.10.

3. What are the characteristic values of a diagonal matrix? Of a triangular matrix? Prove your answers.

4. Prove that if $A$ is nonsingular then the characteristic values of $A^{-1}$ are the reciprocals of the characteristic values of $A$. What can be said about the corresponding characteristic vectors?

5. Show that if $X$ is a characteristic vector of $A$ associated with the value $\lambda$, then for any natural number $k$, $X$ is a characteristic vector of $A^k$ associated with the characteristic value $\lambda^k$. Deduce that if $p$ is any polynomial, then $X$ is a characteristic vector of the matrix $p(A)$, and $p(\lambda)$ is the associated characteristic value.

6. Determine the possible characteristic values of

    (i)  an idempotent matrix,

    (ii)  a nilpotent matrix,

    (iii)  a nonsingular matrix.

7. From the characteristic polynomial of $A$,

$$p(\lambda) = \det (A - \lambda I) = (-1)^n(\lambda - \lambda_1) \ldots (\lambda - \lambda_n),$$

prove that $\det A = \lambda_1\lambda_2 \ldots \lambda_n$. Deduce that $A$ is nonsingular if and only if every characteristic value of $A$ is nonzero.

8. A *companion* matrix is an $n$-by-$n$ matrix $C$ of the form

$$C = \begin{pmatrix} 0 & 0 & . & . & . & 0 & c_1 \\ 1 & 0 & . & . & . & 0 & c_2 \\ 0 & 1 & & & & 0 & c_3 \\ & . & . & & & . & . \\ & . & . & & & . & . \\ 0 & 0 & . & . & . & 1 & c_n \end{pmatrix}.$$

    (i)  Show by induction or otherwise that the characteristic polynomial of $C$ is

$$\det (C - \lambda I) = (-1)^n(\lambda^n - c_n\lambda^{n-1} - \ldots - c_1).$$

    (ii)  Deduce that any polynomial is the characteristic polynomial of some matrix.

9. Prove that if $S = \{\xi_1, \ldots , \xi_k\}$ is a set of characteristic vectors of $\mathbf{T}$ associated respectively with distinct characteristic values $\lambda_1, \ldots , \lambda_k$, then $S$ is linearly independent.

10. Prove Theorem 5.7.

11. Prove that if $\lambda_1, \ldots, \lambda_k$ are distinct characteristic values of **T**, and if $\mathscr{C}_1, \ldots, \mathscr{C}_k$ are the associated characteristic subspaces, then

$$\mathscr{C}_1 \cap (\mathscr{C}_2 + \ldots + \mathscr{C}_k) = [\theta].$$

Deduce that

$$\mathscr{C}_1 + \mathscr{C}_2 + \ldots + \mathscr{C}_k = \mathscr{C}_1 \oplus \mathscr{C}_2 \oplus \ldots \oplus \mathscr{C}_k.$$

12. (i)   Show that $\lambda^2 - (a+d)\lambda + (ad-bc) = 0$ is the characteristic equation of the matrix

$$A = \begin{pmatrix} a & b \\ c & d \end{pmatrix}.$$

(ii)   Also show by matrix computation that

$$A^2 - (a+d)A + (ad-bc)I = Z.$$

## 5.3   Diagonalization of Transformations

From the assertions that were made following Theorem 5.8 it easily follows that if a linear transformation **T** on $\mathscr{V}_n$ has $n$ distinct characteristic values $\lambda_1, \ldots, \lambda_n$, then

$$\mathscr{V}_n = \mathscr{C}_1 \oplus \ldots \oplus \mathscr{C}_n.$$

Each $\mathscr{C}_i$ must then be one-dimensional, and a basis $\{\xi_1, \ldots, \xi_n\}$ of characteristic vectors $\xi_i \in \mathscr{C}_i$ can be chosen for $\mathscr{V}_n$. In this case **T** is *diagonable*; that is, **T** can be represented by a diagonal matrix

$$D = \begin{pmatrix} \lambda_1 & 0 & . & . & . & 0 \\ 0 & \lambda_2 & . & . & . & 0 \\ & . & . & & & . \\ & . & . & & & . \\ & . & . & & & . \\ 0 & 0 & . & . & . & \lambda_n \end{pmatrix}.$$

Even if **T** has only $t$ distinct characteristic values $\lambda_1, \ldots, \lambda_t$, for some $t < n$, **T** is diagonable provided that $\mathscr{V}_n$ is the direct sum of the characteristic subspaces:

$$\mathscr{V}_n = \mathscr{C}_1 \oplus \ldots \oplus \mathscr{C}_t.$$

The essential requirement for diagonability is that a basis of characteristic vectors must exist for $\mathscr{V}_n$.

THEOREM 5.9   An $n$-by-$n$ matrix $A$ is similar to a diagonal matrix $D$ if and only if there exist $n$ linearly independent characteristic vectors of $A$. If $\{X_1, \ldots, X_n\}$ is a set of $n$ linearly independent column vectors such that $AX_j = \lambda_j X_j$ for $j = 1, \ldots, n$ and if $P$ is the matrix having $X_j$ in column $j, j = 1, \ldots, n$, then

$$P^{-1}AP = \text{diag}\ (\lambda_1, \ldots, \lambda_n) = D.$$

PROOF   Suppose that $A$ is similar to a diagonal matrix $D$. Then $A$ and $D$ represent the same linear transformation **T** relative to two bases. But the diagonal form of $D$ shows that each vector $\beta_i$ of the corresponding basis satisfies

$$\mathbf{T}\beta_i = \lambda_i \beta_i.$$

Since **T** has a linearly independent set of $n$ characteristic vectors, so does any matrix that represents **T**. Conversely suppose that $A$ has $n$ linearly independent vectors $X_j$, where $AX_j = \lambda_j X_j$. Form the matrix $P$ whose columns are $X_1, \ldots, X_n$:

$$P = \begin{pmatrix} x_{11} & \cdot & \cdot & \cdot & x_{1n} \\ x_{21} & \cdot & \cdot & \cdot & x_{2n} \\ \cdot & & & & \cdot \\ \cdot & & & & \cdot \\ \cdot & & & & \cdot \\ x_{n1} & \cdot & \cdot & \cdot & x_{nn} \end{pmatrix}.$$

Then from $AX_j = \lambda_j X_j$ we obtain

$$\sum_{k=1}^{n} a_{ik}x_{kj} = \lambda_j x_{ij} \qquad \text{for } i, j = 1, \ldots, n.$$

The left hand expression is the $(i, j)$ entry of $AP$, and the right hand expression is the $(i, j)$ entry of $PD$, so $P^{-1}AP = D$.

Theorem 5.9 provides a necessary and sufficient condition for the diagonability of $A$ and also describes a diagonalizing matrix $P$. The computation needed to carry out the diagonalization is considerable: it requires the complete factorization of the characteristic polynomial of $A$, the complete solution of $t$ systems of $n$ linear equations in $n$ variables (where $t$ is the number of distinct characteristic values of $A$), and the inversion of the $n$-by-$n$ matrix $P$. Some of this work is unnecessary if we are only interested in determining whether $A$ is diagonable, because then we need only to solve the characteristic equation of $A$ and to determine the number of linearly independent characteristic vectors associated with those characteristic values that are repeated roots of the characteristic equation.

**THEOREM 5.10** A sufficient (but not necessary) condition that an $n$-by-$n$ matrix be diagonable is that $A$ have $n$ distinct characteristic values.

PROOF Theorem 5.9 and Exercise 9, Section 5.2.

We shall present two other characterizations of diagonability, each of which has important uses in other aspects of linear algebra, and both of which start from the characteristic polynomial of $A$:

$$p(\lambda) = (-1)^n(\lambda - \lambda_1)^{s_1} \ldots (\lambda - \lambda_t)^{s_t}$$

where $\lambda_1, \ldots, \lambda_t$ are the distinct characteristic values of $A$. If $A$ is diagonable, it is similar to a diagonal matrix $D$ in the following diagonal block form:

$$D = \begin{pmatrix} D_1 & & & \\ & D_2 & & \\ & & \ddots & \\ & & & D_t \end{pmatrix}, \qquad \text{where } D_i = \begin{pmatrix} \lambda_i & & \\ & \ddots & \\ & & \lambda_i \end{pmatrix}.$$

Thus $D_i = \lambda_i I_{s_i}$. The blocks, of course, correspond to the several characteristic subspaces $\mathscr{C}_i$. For each $i = 1, \ldots, t$, let $E_i$ be the $n$-by-$n$ matrix having 1 in each diagonal position in which $\lambda_i$ appears in $D$, and 0 elsewhere:

$$E_i = \begin{pmatrix} Z & & & & \\ & \ddots & & & \\ & & I_{s_i} & & \\ & & & \ddots & \\ & & & & Z \end{pmatrix}.$$

Interpreted as a linear transformation, $E_i$ defines the identity transformation on $\mathscr{C}_i$ and the zero transformation on $\mathscr{C}_j$ when $j \neq i$. In the terminology of Exercise 11 of Section 3.3, such a transformation is a projection. By matrix multiplication it is simple to verify that

$$E_i^2 = E_i \qquad \text{(each } E_i \text{ defines a projection),}$$

$$E_i E_j = Z \quad \text{if } i \neq j \qquad \text{(the projections are orthogonal),}$$

$$\sum_{i=1}^{t} E_i = I \qquad \text{(the projections are supplementary),}$$

$$\sum_{i=1}^{t} \lambda_i E_i = D \qquad \text{(} D \text{ is a linear combination of projections).}$$

But $D = P^{-1}AP$ for $P$ as described in Theorem 5.9. If we let $F_i = PE_iP^{-1}$, then the $F$'s represent the same orthogonal and supplementary projections as the $E$'s, and $A = \sum_{i=1}^{t} \lambda_i F_i$.

**THEOREM 5.11**   An $n$-by-$n$ matrix $A$ is similar to a diagonal matrix $D$ if and only if there exist $t$ distinct scalars $a_1, \ldots, a_t$ and $t$ nonzero idempotent matrices $F_1, \ldots, F_t$ such that

$$F_i F_j = Z \qquad \text{if } i \neq j,$$

$$\sum_{i=1}^{t} F_i = I,$$

$$\sum_{i=1}^{t} a_i F_i = A.$$

PROOF   We have just proved that the diagonability of $A$ implies that such scalars and matrices exist. To prove the converse we shall show that for every vector $X$ and for each $F_j$, $F_jX$ is either $Z$ or a characteristic vector of $A$:

$$A(F_j X) = \left( \sum_{i=1}^{t} a_i F_i \right) F_j X = \left( \sum_{i=1}^{t} a_i F_i F_j \right) X$$

$$= (a_j F_j)X = a_j(F_j X).$$

Hence each $a_j$ is a characteristic value of $A$. Furthermore,

$$X = IX = \sum_{i=1}^{t} F_i X,$$

so every vector is a sum of characteristic vectors. This implies that there are $n$ linearly independent characteristic vectors, so by Theorem 5.9, $A$ is diagonable.

A representation of $A$ as described by Theorem 5.11 is called a *spectral decomposition* of $A$, since the set of scalars $\{a_1, \ldots, a_t\}$ is the spectrum of $A$.

A third criterion for diagonability is related to the Cayley-Hamilton Theorem, one of the most celebrated theorems concerning $n$-by-$n$ matrices.

**THEOREM 5.12** (Cayley-Hamilton). If the characteristic polynomial of an $n$-by-$n$ matrix $A$ is

$$p(\lambda) = (-1)^n(\lambda^n + b_1\lambda^{n-1} + \ldots + b_n),$$

then

$$A^n + b_1 A^{n-1} + \ldots + b_n I = Z.$$

**PROOF** We prove the result only for diagonable matrices; the theorem is true in general, but an extremely simple proof is possible for this special case. If $A$ is diagonable, there exists a basis for $\mathcal{V}_n$ consisting entirely of characteristic vectors of $A$. Let $X_i$ be such a vector, associated with the characteristic value $\lambda_i$. Then by Exercise 5 of Section 5.2, the matrix

$$p(A) = A^n + b_1 A^{n-1} + \ldots + b_n I$$

also has $X_i$ as a characteristic vector, and the associated characteristic value is

$$\lambda_i^n + b_1\lambda_i^{n-1} + \ldots + b_n = 0,$$

since $\lambda_i$ is a characteristic value of $A$. The matrix $p(A)$ carries every vector of a basis into $\theta$, and hence carries every vector of $\mathcal{V}_n$, into $\theta$. Thus $p(A)$ must be the zero matrix.

Since $p(A) = Z$, where $p(x)$ is the characteristic polynomial of $A$, it is conceivable that $q(A) = Z$ for some polynomial $q(x)$ of degree less than $n$.

**DEFINITION 5.4** The *minimal polynomial* $m(x)$ of an $n$-by-$n$ matrix $A$ is the polynomial of lowest degree and having 1 as its leading coefficient such that $m(A) = Z$.

The Cayley-Hamilton Theorem guarantees the existence of $m(x)$, and the requirement that the minimal polynomial have 1 as the coefficient of its highest power makes $m(x)$ unique. Although we shall not take time to pursue this concept, it turns out that the minimal polynomial must be a polynomial divisor of the characteristic polynomial. More precisely, if the characteristic polynomial is

$$p(x) = (-1)^n(x - \lambda_1)^{s_1} \ldots (x - \lambda_t)^{s_t},$$

then

$$m(x) = (x - \lambda_1)^{r_1} \ldots (x - \lambda_t)^{r_t},$$

where $1 \le r_i \le s_i$ for $i = 1, \ldots, t$. Every characteristic value of $A$ is a zero of the minimal polynomial of $A$. Furthermore, if $A$ is diagonable, $r_i = 1$ for $i = 1, \ldots, t$. In a sense, the extent to which $m(x)$ has multiple roots is a measure of the extent to which $A$ fails to be diagonable.

**THEOREM 5.13** Let $A$ be an $n$-by-$n$ matrix whose distinct characteristic values are $\lambda_1, \ldots, \lambda_t$. $A$ is diagonable if and only if

$$(A - \lambda_1 I) \ldots (A - \lambda_t I) = Z.$$

This theorem provides a third criterion for diagonability, one that can be applied computationally as soon as a complete set of characteristic values is known.

Finally, we say a few words about the more difficult problem that we have not resolved. For a matrix or transformation that is not diagonable, what is a simple canonical form for similarity? The answer is that we can come very close to a diagonal form. Any $n$-by-$n$ matrix $A$ is similar to a matrix of the form

$$J = \begin{pmatrix} \lambda_1 & & & & \\ * & \lambda_2 & & & \\ & * & \ddots & & \\ & & & \ddots & \\ & & & * & \lambda_n \end{pmatrix}$$

in which the diagonal entries are the characteristic values of $A$, the subdiagonal entries are either 0 or 1, and every other entry is 0. Any square matrix, therefore, is similar to the sum of a diagonal matrix and a nilpotent matrix. Furthermore, the exact pattern of 0's and 1's on the subdiagonal is completely determined by the geometric behavior of the linear transformation represented by $A$.

## EXERCISES 5.3

1. For each of the following matrices determine the characteristic values and corresponding characteristic vectors; if a given matrix is similar to a diagonal matrix, find a diagonalizing matrix $P$ and verify that $P^{-1}AP$ is diagonal.

(i) $\begin{pmatrix} -2 & 3 & -1 \\ -6 & 7 & -2 \\ -9 & 9 & -2 \end{pmatrix}$.

(iv) $\begin{pmatrix} 1 & 0 & 0 \\ 1 & 0 & -2 \\ -1 & 1 & -3 \end{pmatrix}$.

(ii) $\begin{pmatrix} 7 & 4 & -4 \\ 4 & 7 & -4 \\ -1 & -1 & 4 \end{pmatrix}$.

(v) $\begin{pmatrix} 1 & 3 & 0 \\ 0 & -2 & 0 \\ 0 & 6 & 1 \end{pmatrix}$.

(iii) $\begin{pmatrix} 2 & 10 & 5 \\ -2 & -4 & -4 \\ 3 & 5 & 6 \end{pmatrix}$.

(vi) $\begin{pmatrix} 2 & 2 & 1 \\ 1 & 3 & 1 \\ 1 & 2 & 2 \end{pmatrix}$.

2. Which matrices of Exercise 1 of Section 5.2 are diagonable, and why? Without computing a diagonalizing matrix $P$, write a diagonal form of each matrix that is diagonable.

3. Which of the following matrices are diagonable, and why? (Exercise 8 of Section 5.2 will simplify finding the characteristic polynomial for three of these matrices.)

(i) $\begin{pmatrix} 2 & 0 & 0 \\ 0 & 1 & 0 \\ 0 & 2 & 2 \end{pmatrix}$.

(iii) $\begin{pmatrix} 0 & 0 & 0 \\ 1 & 0 & -9 \\ 0 & 1 & 6 \end{pmatrix}$.

(ii) $\begin{pmatrix} 0 & 0 & -1 \\ 1 & 0 & -3 \\ 0 & 1 & -3 \end{pmatrix}$.

(iv) $\begin{pmatrix} 0 & 0 & 0 & 0 \\ 1 & 0 & 0 & -4 \\ 0 & 1 & 0 & 4 \\ 0 & 0 & 1 & 1 \end{pmatrix}$.

4. Show that $A$ and $B$ are similar, where

$$A = \begin{pmatrix} 1 & 1 & . & . & . & 1 \\ 1 & 1 & . & . & . & 1 \\ . & . & & & & . \\ . & . & & & & . \\ . & . & & & & . \\ 1 & 1 & . & . & . & 1 \end{pmatrix}, \quad B = \begin{pmatrix} n & 0 & . & . & . & 0 \\ 0 & 0 & . & . & . & 0 \\ . & . & & & & . \\ . & . & & & & . \\ . & . & & & & . \\ 0 & 0 & . & . & . & 0 \end{pmatrix}.$$

5. Given $n$-by-$n$ matrices $E_1, \ldots, E_t$ such that $E_iE_j = Z$ if $i \neq j$ and $E_1 + \ldots + E_t = I$.

(i)  Prove that each $E_i$ is idempotent.

(ii)  For $i = 1, \ldots, t$ let $F_i = PE_iP^{-1}$ for some nonsingular matrix $P$. Show by matrix algebra that each $F_i$ is idempotent, and show that the set $\{F_1, \ldots, F_t\}$ is orthogonal and supplementary.

6. Write a spectral decomposition for each of the matrices in Exercises 1(ii) and 3(i).

7. Use the Cayley-Hamilton Theorem to prove that if $A$ is a singular $n$-by-$n$ matrix, then

(i)  $A^2$ is proportional to $A$ if $n = 1$ or 2, but

(ii)  $A^2$ need not be proportional to $A$ if $n > 2$.

8. (i)  Determine all real two-by-two matrices $A$ that satisfy $A^2 = -I$.

(ii)  Prove that no real three-by-three matrix $A$ satisfies $A^2 = -I$.

9. Prove that an $n$-by-$n$ matrix is singular if and only if the constant term of its minimal polynomial is zero.

10. Find the minimal polynomial of each of the matrices specified below, and verify your result by computation:

(i)  Exercise 1(iv) of Section 4.3.
(ii)  Exercise 1(ii) of this section.
(iii)  Exercise 1(vi) of this section.
(iv)  Exercise 3(ii) of this section.

## 5.4  Quadratic Forms

In these final two sections we return to a study of quadratic forms on $E_n$, begun in Section 1.11 for the case $n = 3$. Again the basic questions are: how is a quadratic form represented by a matrix; how is the representation altered by a change of coordinates; and how can coordinates be chosen to obtain a simple representation?

**DEFINITION 5.5**   A real quadratic form in $x_1, \ldots, x_n$ is an expression of the form

$$\sum_{i=1}^{n} \sum_{j=1}^{n} a_{ij}x_ix_j.$$

The form is said to be *positive definite* if and only if

$$\sum_{i=1}^{n} \sum_{j=1}^{n} a_{ij}x_ix_j \geq 0 \qquad \text{for all } x_1, \ldots, x_n,$$

with equality holding only when $x_1 = \ldots = x_n = 0$.

Relative to any basis $\{\alpha_1, \ldots, \alpha_n\}$, let $\xi = \sum_{i=1}^{n} x_i\alpha_i$. Then the form assigns a real number $q(\xi)$ to $\xi$, and therefore $q$ can be regarded as a quadratic function from $E_n$ to $R$:

$$q(\xi) = \sum_{i=1}^{n} \sum_{j=1}^{n} x_i a_{ij} x_j.$$

In matrix notation this expression becomes

$$q(\xi) = X^t A X,$$

where $A = (a_{ij})$. Since for $i \neq j$ a quadratic form contains the sum

$$a_{ij}x_ix_j + a_{ji}x_jx_i = (a_{ij} + a_{ji})x_ix_j,$$

we can symmetrize the form by agreeing that each of the coefficients of $x_ix_j$ and $x_jx_i$ will be $\frac{1}{2}(a_{ij} + a_{ji})$.

**DEFINITION 5.6** Let $A$ be an $n$-by-$n$ matrix $(a_{ij})$.
(a)  $A$ is said to be *symmetric* if and only if $A^t = A$.
(b)  $A$ is said to be *skew symmetric* if and only if $A^t = -A$.

As an exercise you may verify that $A + A^t$ is symmetric and $A - A^t$ is skew, for any $n$-by-$n$ matrix $A$. Hence $A$ is the sum of a symmetric matrix, $\frac{1}{2}(A + A^t)$, and a skew matrix, $\frac{1}{2}(A - A^t)$. Furthermore, you may show that if $A = S + K$, where $S$ is symmetric and $K$ is skew, then for any vector $X$, $X^t K X = 0$, and so

$$X^t A X = X^t S X + X^t K X = X^t S X.$$

This observation justifies our agreement to symmetrize a given quadratic form. Henceforth we shall assume that this has been done and that the form is represented by a symmetric matrix.

Now, what happens to this representation under a change from the

$\alpha$-basis to a $\gamma$-basis? If we express the new basis vectors in terms of the old, we obtain a nonsingular matrix $P = (p_{ij})$ from the relation

$$\gamma_j = \sum_{i=1}^n p_{ij}\alpha_i.$$

Then

$$\xi = \sum_{i=1}^n x_i\alpha_i = \sum_{j=1}^n y_j\gamma_j$$
$$= \sum_{j=1}^n y_j\left(\sum_{i=1}^n p_{ij}\alpha_i\right) = \sum_{i=1}^n \left(\sum_{j=1}^n p_{ij}y_j\right)\alpha_i,$$

so

$$x_i = \sum_{j=1}^n p_{ij}y_j, \text{ or } X = PY.$$

Therefore

$$q(\xi) = X^t AX = (PY)^t A(PY) = Y^t(P^t AP)Y.$$

Since $\xi$ is represented by $Y$ in the new coordinate system, the quadratic form $q$ is represented by the matrix $C = P^t AP$. Note that $C^t = (P^t AP)^t = P^t A^t(P)^t = P^t AP = C$, since $A$ is symmetric. We have proved the following theorem.

**THEOREM 5.14**   Let $q$ be a quadratic form that is represented relative to a given basis for $E_n$ by the symmetric matrix $A$. Relative to a new basis, $q$ is represented by the symmetric matrix $C = P^t AP$, where $P$ is the nonsingular matrix whose columns represent the new basis vectors in terms of the original basis.

Note the resemblance of this result to the corresponding analysis of the matrix representation of linear transformations, given by specializing Theorem 5.1 to the case where $P = Q$. Matrices $A$ and $P^{-1}AP$ that represent the same linear transformation were said to be similar. Matrices $A$ and $P^t AP$ that represent the same quadratic form are said to be congruent.

**DEFINITION 5.7**   A real $n$-by-$n$ matrix $C$ is said to be *congruent* to a real $n$-by-$n$ matrix $A$ if and only if there exists a nonsingular matrix $P$ such that

$$C = P^t AP.$$

**THEOREM 5.15** Congruence is an equivalence relation of the set $\mathcal{M}_{n \times n}$ of all real $n$-by-$n$ matrices. Furthermore, any matrix that is congruent to a symmetric matrix is symmetric, and any matrix that is congruent to a skew matrix is skew.

**PROOF** Exercise.

Since a given quadratic form can be represented by any matrix of a corresponding congruence class of symmetric matrices, we are interested in obtaining a canonical form for congruence of real symmetric matrices. The key to the investigation is to recall that any nonsingular matrix $P$ is a product of elementary matrices, $E_1 E_2 \ldots E_k$:

$$P^t A P = (E_1 E_2 \ldots E_k)^t A (E_1 E_2 \ldots E_k)$$

$$= E_k^t \ldots E_1^t A E_1 \ldots E_k.$$

Also $E_i^t$ performs on the rows of $A$ the same elementary operations as $E_i$ performs on the columns. Hence, starting with a symmetric matrix $A$, we pass to a symmetric matrix that is congruent to $A$ by performing a row operation and the corresponding column operation. The operations that we choose are motivated by the same objectives as in Gaussian elimination.

**THEOREM 5.16** Each real symmetric $n$-by-$n$ matrix $A$ of rank $r$ is congruent to one and only one diagonal matrix having 1 in the first $p$ diagonal positions, $-1$ in the next $r - p$ diagonal positions, and 0 elsewhere. The number $p$ is determined by $A$.

**PROOF** If $A = Z$, then $r = p = 0$, and there is nothing to do. Otherwise, our first objective is to produce a nonzero entry in the $(1, 1)$ position. If all the diagonal elements of $A$ are zero, let $a_{ij} \neq 0$ for some $i \neq j$. Add row $i$ to row $j$ and column $i$ to column $j$ to produce a matrix $B$ that is congruent to $A$ and has $b_{jj} = a_{ij} + a_{ji} \neq 0$. If some diagonal element of $A$ is nonzero, that step is unnecessary.

In either case $A$ is congruent to a matrix $B$ having $b_{jj} \neq 0$. Permute rows 1 and $j$ and columns 1 and $j$ of $B$ to obtain a symmetric matrix $C$ congruent to $B$ and $A$ and having $c_{11} = b_{jj} \neq 0$. Then row operations can be used to produce zeros in the other positions of column 1, and the corresponding column operations produce zeros in the first row to yield a matrix of the form

$$P^tAP = \begin{pmatrix} d_{11} & 0 & . & . & . & 0 \\ 0 & & & & & \\ . & & & A_1 & & \\ . & & & & & \\ . & & & & & \\ 0 & & & & & \end{pmatrix},$$

where $d_{11} = c_{11} \neq 0$ and where $P$ denotes the product of the elementary matrices used to perform the elementary *column* operations used in this process. If $A_1 = Z$ we are through. Otherwise the process is repeated on the rows and columns of $A_1$. After $r$ applications of the entire process we have a matrix $D$ congruent to $A$ and of the form

$$D = \begin{pmatrix} d_1 & & & & Z \\ & . & & & \\ & & . & & \\ & & & d_r & \\ & Z & & & Z \end{pmatrix}, \quad d_i \neq 0 \quad \text{for } i = 1, \ldots, r.$$

By applying a single row and column permutation to $D$ we can permute any two diagonal elements, so if $p$ of the scalars $d_i$ are positive, $A$ is congruent to a diagonal matrix

$$\text{diag } (e_1, \ldots, e_p, e_{p+1}, \ldots, e_r, 0, \ldots, 0),$$

where $e_i > 0$ for $i = 1, \ldots, p$ and $e_i < 0$ for $i = p + 1, \ldots, r$. Multiply row $i$ and column $i$ by $e_i^{-1/2}$ for $i = 1, \ldots, p$, and by $(-e_i)^{-1/2}$ for $i = p + 1, \ldots, r$ to obtain a matrix $F$ congruent to $A$ and of the diagonal block form

$$F = \begin{pmatrix} I_p & & \\ & -I_{r-p} & \\ & & Z \end{pmatrix}.$$

To show that $F$ is the only matrix of this form congruent to $A$, suppose that $A$ is congruent to a matrix $G$ of the diagonal block form

$$G = \begin{pmatrix} I_q & & \\ & -I_{s-q} & \\ & & Z \end{pmatrix}.$$

Then $r = s$ since congruent matrices have the same rank. Also $F$ and $G$ represent the same quadratic form $q$ relative to different bases. Let

$$\xi = \sum_{i=1}^{n} x_i \alpha_i = \sum_{i=1}^{n} y_i \beta_i.$$

Then

$$q(\xi) = x_1^2 + \ldots + x_p^2 - x_{p+1}^2 - \ldots - x_r^2,$$

$$q(\xi) = y_1^2 + \ldots + y_q^2 - y_{q+1}^2 - \ldots - y_r^2.$$

If $p > q$, let $\mathcal{M} = [\alpha_1, \ldots, \alpha_p]$ and $\mathcal{N} = [\beta_{q+1}, \ldots, \beta_n]$. Then $q(\mu) > 0$ for every nonzero $\mu \in \mathcal{M}$, and $q(\nu) \leq 0$ for every $\nu \in \mathcal{N}$. Hence $\mathcal{M} \cap \mathcal{N} = [\theta]$. But

$$\dim \mathcal{M} + \dim \mathcal{N} = p + (n - q) = n + (p - q) > n,$$

$$\dim \mathcal{M} + \dim \mathcal{N} = \dim (\mathcal{M} + \mathcal{N}) + \dim (\mathcal{M} \cap \mathcal{N})$$

$$\leq n + 0.$$

These two inequalities are a contradiction, so $p \leq q$. By reversing the roles, we can obtain $q \leq p$, so $p = q$ and the proof is complete.

The fact that the matrix $F$ of Theorem 5.16 is canonical for congruence of real symmetric matrices means that the numbers $p$ and $r$ are *uniquely* determined for any real symmetric matrix $A$, and that two such matrices are congruent if and only if they determine the same numbers $p$ and $r$.

**DEFINITION 5.8**   The *signature* $s$ of a real symmetric matrix $A$ is defined by

$$s = p - (r - p) = 2p - r;$$

$s$ denotes the number of 1's diminished by the number of $-1$'s along the diagonal of the canonical matrix which is congruent to $A$.

**THEOREM 5.17**   Two real symmetric matrices are congruent if and only if they have the same rank and the same signature.
PROOF   This is immediate from Theorem 5.16.

For a final theorem, we can group the three statements on the next page.

**THEOREM 5.18** The following statements are equivalent:

(a)  A quadratic form $q(\xi)$ is positive definite on $E_n$,

(b)  Any matrix $A$ that represents $q$ has rank $n$ and signature $n$,

(c)  $A = Q^t Q$ for some real nonsingular matrix $Q$.

PROOF   Exercise.

## EXERCISES 5.4

1. Show that any quadratic function $q$ satisfies

$$q(\xi + \eta) + q(\xi - \eta) = 2q(\xi) + 2q(\eta).$$

Explain why this relation is called the parallelogram identity.

2. Represent each of the following quadratic forms by a real symmetric matrix, and determine the rank and signature of each:

(i)  $x_1^2 - 2x_1 x_3 + 2x_2^2 + 4x_2 x_3 + 6x_3^2$.

(ii)  $16x_1 x_2 - x_3^2$.

(iii)  $3x_1^2 + 4x_1 x_2 + 8x_1 x_3 + 4x_2 x_3 + 3x_3^2$.

3. Verify the assertions made in the paragraph following Definition 5.6.

4. Prove that $A^2$ is symmetric if either $A$ is symmetric or $A$ is skew.

5. If $A$ and $B$ are both symmetric, prove that

(i)  $A + B$ is symmetric,

(ii)  $AB$ is symmetric if and only if $A$ and $B$ commute.

6. If $A$ and $B$ are skew, prove that $A + B$ is skew.

7. Decompose the matrix $A$ into the sum $S + K$ of a symmetric matrix and a skew matrix, where

$$A = \begin{pmatrix} 2 & 3 & -4 \\ 1 & 0 & 5 \\ 0 & -3 & 1 \end{pmatrix}.$$

8. Let $A$ be the matrix of Exercise 7, and let $X^t = (x_1, x_2, x_3)$.

(i)  Calculate $X^t A X$.

(ii)  Let $A = S + K$, where $S$ is symmetric and $K$ is skew; calculate $X^t S X$ and $X^t K X$.

(iii)  Prove in general that, if $K$ is a real skew $n$-by-$n$ matrix and $X$ a real $n$-by-1 matrix, then $X^t K X = 0$.

9. Prove Theorem 5.15.

10. Prove Theorem 5.18.

11. Determine the rank and signature of each of the following real symmetric matrices.

(i)   $A = \begin{pmatrix} 4 & -5 & -2 \\ -5 & 4 & -2 \\ -2 & -2 & -8 \end{pmatrix}$.

(iv)   $D = \begin{pmatrix} -1 & -3 & 1 \\ -3 & -4 & 2 \\ 1 & 2 & -2 \end{pmatrix}$.

(ii)   $B = \begin{pmatrix} 5 & -2 & -5 \\ -2 & 1 & 2 \\ -5 & 2 & 5 \end{pmatrix}$.

(v)   $E = \begin{pmatrix} 1 & 2 & 5 \\ 2 & 2 & 2 \\ 5 & 2 & -7 \end{pmatrix}$.

(iii)   $C = \begin{pmatrix} 5 & -3 & -1 \\ -3 & 3 & 1 \\ -1 & 1 & -1 \end{pmatrix}$.

(vi)   $F = \begin{pmatrix} -1 & 1 & 3 \\ 1 & 1 & 1 \\ 3 & 1 & -1 \end{pmatrix}$.

## 5.5   Diagonalization of Quadratic Forms

With physical and geometric applications in mind, we are particularly interested in simplifying a quadratic form by using only rigid motions to change coordinates; that is, we shall deliberately restrict our attention to linear transformations that preserve the metric properties of $E_n$. A linear rigid motion of $E_n$ is an orthogonal transformation **T**, discussed in Section 3.3, which carries a normal orthogonal basis into a normal orthogonal basis. Relative to such a basis, **T** is represented by an orthogonal matrix $P$, described in Section 3.6. The column vectors of $P$ are of unit length and mutually orthogonal:

$$\sum_{i=1}^{n} p_{ij}p_{ik} = \delta_{jk}.$$

Hence $P^{-1} = P^t$. If a quadratic form is represented by a real symmetric matrix $A$, and if an orthogonal change of basis is made, the form is then represented by

$$C = P^tAP = P^{-1}AP.$$

Hence $C$ is simultaneously congruent to $A$ and similar to $A$. Since characteristic values and vectors played a central role in analyzing similarity, we resume our consideration of quadratic forms by examining the characteristic values of real symmetric matrices. We shall discover that such matrices have remarkable properties.

**THEOREM 5.19**   The characteristic polynomial of a real symmetric matrix $A$ is a product of real linear factors. Every characteristic value is real.

**PROOF**   For the moment we consider $A$ as a real matrix over the complex field, where we know that the characteristic polynomial factors into linear factors

$$p(x) = (-1)^n(x - \lambda_1)^{s_1} \ldots (x - \lambda_t)^{s_t}.$$

Let $\lambda$ be any characteristic value and $X$ any associated characteristic vector, where $\lambda$ and the components of $X$ may be complex numbers:

$$AX = \lambda X,$$

$$\bar{X}^t A X = \lambda \bar{X}^t X,$$

where $\bar{X}^t$ is the row vector whose components are the complex conjugates of the components of $X$. Then,

$$\bar{X}^t X = \sum_{i=1}^{n} \bar{x}_i x_i$$

is a positive real number since the product of a complex number and its conjugate is a nonnegative real number. (See Exercise 1.) Likewise, $\bar{X}^t A X$ is a scalar $c$, conceivably complex, so

$$c = \bar{X}^t A X = (\bar{X}^t A X)^t = X^t A \bar{X},$$

since $A^t = A$. Also

$$c = \sum_{i=1}^{n} \sum_{j=1}^{n} \bar{x}_i a_{ij} x_j,$$

$$\bar{c} = \sum_{i=1}^{n} \sum_{j=1}^{n} x_i a_{ij} \bar{x}_j,$$

since $\overline{u + v} = \bar{u} + \bar{v}$, $\overline{uv} = \bar{u}\bar{v}$, and $\bar{a}_{ij} = a_{ij}$ because $A$ is real. Hence

$$\bar{c} = X^t A \bar{X} = c,$$

so $c$ is real. Thus the product of $\lambda$ and the positive real number $\bar{X}^t X$ is real, so $\lambda$ is real. Since $X$ is a solution of the system of linear equations

$$(A - \lambda I)X = Z,$$

having real coefficients, $X$ is real.

It is also interesting to note that for a real symmetric matrix any characteristic vectors in $E_n$ associated with distinct characteristic values are perpendicular.

**THEOREM 5.20**   Let $X_1$ and $X_2$ be characteristic vectors associated with distinct characteristic values $\lambda_1$ and $\lambda_2$ of a real symmetric matrix $A$. Then $X_1{}^t X_2 = 0$.

**PROOF**   Let $AX_1 = \lambda_1 X_1$ and $AX_2 = \lambda_2 X_2$. Then

$$X_1{}^t(AX_2) = X_1{}^t(\lambda_2 X_2) = \lambda_2 X_1{}^t X_2$$

$$= (X_1{}^t A)X_2 = (AX_1)^t X_2$$

$$= (\lambda_1 X_1)^t X_2 = \lambda_1 X_1{}^t X_2.$$

Since $(\lambda_1 - \lambda_2)X_1{}^t X_2 = 0$ and $\lambda_1 \neq \lambda_2$, $X_1{}^t X_2 = 0$. In a Euclidean space $X_1{}^t X_2 = \langle \xi_1, \xi_2 \rangle$ relative to a normal orthogonal basis (Theorem 2.21), so $\xi_1$ and $\xi_2$ are orthogonal if they are characteristic vectors associated with different characteristic values of a real symmetric matrix.

The next result is the matrix form of the important Principal Axes Theorem, which shows that any quadratic form can be reduced by an orthogonal transformation to a sum of squares. In particular, the axes of any quadric surface are orthogonal, and a suitable rigid motion of the axes of any rectangular coordinate system in $E_3$ will align the new axes with the surface. You will recall that an analogous result holds for conic sections in $E_2$.

**THEOREM 5.21**   (Principal Axes Theorem). If $A$ is a real symmetric $n$-by-$n$ matrix, there exists an orthogonal matrix $P$ such that $P^{-1}AP$ is diagonal, with the characteristic values of $A$ on the diagonal.

**PROOF**   Let $\{\alpha_1, \ldots, \alpha_n\}$ be any normal orthogonal basis for $E_n$, and let $\mathbf{T}$ be the linear transformation represented by $A$ relative to that basis. If $\xi$ is any characteristic vector of $\mathbf{T}$ associated with $\lambda_1$, then $\beta_1 = \|\xi\|^{-1}\xi$ is a characteristic vector of unit length, also associated with $\lambda_1$. Extend to a normal orthogonal basis $\{\beta_1, \beta_2, \ldots, \beta_n\}$. Relative to this basis $\mathbf{T}$ is represented by the matrix

$$B = R^{-1}AR = R^t AR,$$

since $R$ is orthogonal. $B$ is real and symmetric,

$$B = \left(\begin{array}{c|c} \lambda_1 & Z \\ \hline Z & B_1 \end{array}\right),$$

so $B_1$ is real and symmetric. The process can be repeated on the T-invariant space $[\beta_2, \ldots, \beta_n]$, selecting a unit characteristic vector $\gamma_2$ associated with $\lambda_2$. Letting $\gamma_1 = \beta_1$, we can extend the normal orthogonal set $\{\gamma_1, \gamma_2\}$ to a normal orthogonal basis $\{\gamma_1, \ldots, \gamma_n\}$, relative to which **T** is represented by

$$C = S^{-1}BS = S^t BS = \left(\begin{array}{cc|c} \lambda_1 & 0 & \\ 0 & \lambda_2 & Z \\ \hline Z & & C_1 \end{array}\right).$$

Since $R$ and $S$ are orthogonal, so is $RS$, and

$$C = S^{-1}(R^{-1}AR)S = (RS)^{-1}A(RS)$$

$$= (RS)^t A(RS).$$

After $n$ steps we obtain a diagonal matrix $D = \text{diag}(\lambda_1, \ldots, \lambda_n)$ and an orthogonal matrix $P$ such that $D = P^{-1}AP = P^t AP$.

The fact that $D$ is simultaneously similar to and congruent to $A$ emphasizes that these two equivalence relations are identical when the transforming matrix $P$ is orthogonal. Of course in the reduction of $A$ to diagonal form we could have selected the characteristic values in any order. Hence, by using the positive characteristic values first, then the negative characteristic values, we can use orthogonal changes of bases to transform the real symmetric matrix $A$ into $D = \text{diag}(\lambda_1, \ldots, \lambda_n)$. But this is as far as similarity carries us. To obtain the canonical form for congruence given in Theorem 5.16 we must use a nonorthogonal transformation amounting to a change of scale along each axis.

In the process we have discovered another way to determine the rank and signature of a real symmetric matrix: if the $n$ (not necessarily distinct) characteristic values of $A$ are listed with all the positive values grouped first, then the negative values, and then the zeros,

$$(\lambda_1, \ldots, \lambda_p)(\lambda_{p+1}, \ldots, \lambda_r)(\lambda_{r+1}, \ldots, \lambda_n),$$

then $r$ is the rank of $A$ and $2p - r$ is the signature of $A$.

Our final result is of importance in such applications as vibration problems in dynamics. Given two quadratic forms, one of which is positive definite, there exists a single change of coordinates in $E_n$ that diagonalizes both forms. This result is easily visualized in terms of central quadric surfaces in $E_3$. Each such surface determines a quadratic form, and ellipsoids correspond to positive definite forms. Given a central ellipsoid and another central quadric surface, first align the coordinate axes with the axes of the ellipsoid, an orthogonal transformation. Then change scale along the axes of the ellipsoid, deforming it into a sphere, for which any direction is a principal axis. Then rotate axes again to align the coordinate axes with the principal axes of the second surface.

**THEOREM 5.22**   Let $A$ and $B$ be real symmetric matrices, and let $A$ be positive definite. There exists a nonsingular matrix $P$ such that

$$P^tAP = I,$$

$$P^tBP = D, \quad \text{where } D \text{ is diagonal.}$$

**PROOF**   Since $A$ is positive definite, it is congruent to $I$;

$$Q^tAQ = I \text{ for some nonsingular } Q.$$

Then for *any* orthogonal matrix $R$, $R^t = R^{-1}$, and

$$R^t(Q^tAQ)R = R^{-1}IR,$$

$$(QR)^tA(QR) = I.$$

Also $Q^tBQ$ is real and symmetric, so by the Principal Axes Theorem we can choose an orthogonal $R$ such that

$$R^t(Q^tBQ)R = D.$$

Let $P = QR$, so that $P^tAP = I$, and $P^tBP = D$.

## EXERCISES 5.5

1. Operations for complex numbers are defined as follows, where $a, b, c,$ and $d$ are real:

   Sum: $(a + ib) + (c + id) = (a + c) + i(b + d)$.

   Product: $(a + ib)(c + id) = (ac - bd) + i(ad + bc)$.

   Conjugate: $\overline{a + ib} = a - ib$.

   Magnitude: $|a + ib| = \sqrt{a^2 + b^2}$.

Show that if $u$ and $v$ are complex numbers ($u = a + ib$ and $v = c + id$), then

(i) $\overline{u + v} = \bar{u} + \bar{v}$,

(ii) $\overline{uv}, = \bar{u}\bar{v}$

(iii) $\bar{\bar{u}} = u$,

(iv) $u\bar{u} = |u|^2$,

(v) $u + \bar{u}$ is real,

(vi) $|u|$ is real and nonnegative,

(vii) $|uv| = |u|\,|v|$,

(viii) $|u + v| \le |u| + |v|$.

2. Given the real quadratic form $ax_1^2 + 2bx_1x_2 + cx_2^2$.

(i) Prove that the form is positive definite if and only if $a > 0$ and $b^2 - ac < 0$.

(ii) Show that the central conic $ax^2 + 2bxy + cy^2 = 1$ is an ellipse or hyperbola, depending upon whether the quadratic form on the left has $r = 2$ and $s = 2$, or $r = 2$ and $s = 0$.

3. As we know, any quadratic form in three variables can be reduced by an orthogonal change of coordinates to the form

$$\lambda_1 x_1^2 + \lambda_2 x_2^2 + \lambda_3 x_3^2,$$

where the $\lambda_i$ are real. Hence any centrally symmetric quadric surface has an equation of the form

$$\lambda_1 x_1^2 + \lambda_2 x_2^2 + \lambda_3 x_3^2 = 1$$

in a suitable rectangular coordinate system. Use the rank and signature of the quadratic form to classify all possible types of centrally symmetric quadric surfaces, and identify each type by means of a sketch.

4. The Taylor expansion of a function $f$ of two variables at $(a, b)$ is expressed in terms of the partial derivatives of $f$ by

$$f(a + h, b + k) = f(a, b) + hf_x(a, b) + kf_y(a, b)$$

$$+ \frac{1}{2}[h^2 f_{xx}(a, b) + 2hk f_{xy}(a, b) + k^2 f_{yy}(a, b)] + \dots,$$

provided that $f_{xy} = \dfrac{\partial}{\partial y}\left(\dfrac{\partial f}{\partial x}\right) = \dfrac{\partial}{\partial x}\left(\dfrac{\partial f}{\partial y}\right) = f_{yx}$. If $f_x(a, b) = 0 = f_y(a, b)$, then $(a, b)$ is a critical point for maximum or minimum. The term in brackets is a quadratic form in $h$ and $k$; if the form has rank 2, it determines whether $f$ has a maximum or minimum or neither at $(a, b)$. Assuming $r = 2$, show that

(i) $f(a, b)$ is a relative maximum if $s = -2$,

(ii) $f(a, b)$ is a relative minimum if $s = 2$,

(iii)  $f(a, b)$ is neither maximum nor minimum otherwise.

5. Prove that if $P$ is an orthogonal matrix, then

(i)   $\det P = \pm 1$.

(ii)  $|\lambda| = 1$ for each characteristic value $\lambda$.

6. Prove that if $K$ is a skew matrix such that $I + K$ is nonsingular, then $(I - K)(I + K)^{-1}$ is orthogonal.

# Greek Alphabet

| Name | Lower Case | Capital |
|------|------------|---------|
| alpha | $\alpha$ | A |
| beta | $\beta$ | B |
| gamma | $\gamma$ | $\Gamma$ |
| delta | $\delta$ | $\Delta$ |
| epsilon | $\epsilon$ | E |
| zeta | $\zeta$ | Z |
| eta | $\eta$ | H |
| theta | $\theta$ | $\Theta$ |
| iota | $\iota$ | I |
| kappa | $\kappa$ | K |
| lambda | $\lambda$ | $\Lambda$ |
| mu | $\mu$ | M |
| nu | $\nu$ | N |
| xi | $\xi$ | $\Xi$ |
| omicron | $o$ | O |
| pi | $\pi$ | $\Pi$ |
| rho | $\rho$ | P |
| sigma | $\sigma$ | $\Sigma$ |
| tau | $\tau$ | T |
| upsilon | $\upsilon$ | $\Upsilon$ |
| phi | $\phi$ | $\Phi$ |
| chi | $\chi$ | X |
| psi | $\psi$ | $\Psi$ |
| omega | $\omega$ | $\Omega$ |

# Suggestions and Answers
## for Exercises

# 1

**EXERCISES 1.1** *pages 6–8*

1.  (i)  (1, 2);                 (v)  (−2, 1);
    (ii)  (−1, −2);            (vi)  (1, 2);
    (iii)  (1, 2);               (vii)  (2, 4);
    (iv)  (2, 1);                (i), (iii), (vi) are congruent.

2.  (i)  The line of slope 3/2 through (0, 0).
    (ii)  The line of slope 3/2 through (−1, 1).
    (iii)  The half line of slope 3/2 downward from (1, 4).
    (iv)  The line segment between (2, 3) and (−1, 1).

3.  Let $\xi = (x, y)$ and $\xi = \beta + t\alpha$.
$$x = b_1 + ta_1;$$
$$y = b_2 + ta_2;$$
$$y - b_2 = \frac{a_2}{a_1}(x - b_1), \quad \text{if } a_1 \neq 0.$$

4.  $\xi = t\alpha + (1 - t)\beta$ is the line segment between $A$ and $B$;

$$x = ta_1 + (1 - t)b_1;$$
$$y = ta_2 + (1 - t)b_2;$$
$$y - b_2 = \frac{a_2 - b_2}{a_1 - b_1}(x - b_1), \qquad \text{if } a_1 \neq b_1.$$

5. (i)  Midpoint of the line segment from $\beta$ to $\alpha$.

   (ii)  A point on the segment from $\beta$ through $\alpha$, beyond $\alpha$.

   (iii)  All points on the line from $\alpha$ through $\beta$, beyond $\beta$.

6. (i)  $(\alpha \cdot \alpha)^{1/2} = \sqrt{a_1^2 + a_2^2} = \|\alpha\|$

   (ii)  Consider the slopes of the lines along $\alpha$ and $\beta$.

   (iii)  By the law of cosines,

$$\|\beta - \alpha\|^2 = \|\alpha\|^2 + \|\beta\|^2 - 2\|\alpha\|\,\|\beta\|\cos\Psi$$
$$= (\beta - \alpha) \cdot (\beta - \alpha)$$
$$= \|\beta\|^2 + \|\alpha\|^2 - 2\alpha \cdot \beta.$$

   (iv)  The perpendicular projection of $\beta$ on $\alpha$ is $k\dfrac{\alpha}{\|\alpha\|}$ where $k = \|\beta\|\cos\Psi$; the perpendicular projection of $\beta$ on $\alpha$ is therefore

$$\frac{\|\alpha\|\,\|\beta\|\cos\Psi}{\|\alpha\|^2}\,\alpha = \frac{\alpha \cdot \beta}{\alpha \cdot \alpha}\,\alpha.$$

   (v)  Area of the parallelogram is

$$\|\beta\|\,\|\alpha\|\sin\Psi = \|\alpha\|\,\|\beta\|\,(1 - \cos^2\Psi)^{1/2}$$
$$= [(\alpha \cdot \alpha)(\beta \cdot \beta) - (\alpha \cdot \beta)^2]^{1/2}.$$

   (vi)  Area $= [(a_1^2 + a_2^2)(b_1^2 + b_2^2) - (a_1b_1 + a_2b_2)^2]^{1/2}$
$$= [a_1^2b_2^2 + a_2^2b_1^2 - 2a_1b_1a_2b_2]^{1/2} = |(a_1b_2 - a_2b_1)|.$$

7. (i)  $(-2, 1) \cdot (1, 2) = 0$, hence perpendicular.

   (ii)  $\|(2, 1)\| = \sqrt{5} = \|(1, 2)\|$.

   (iii)  Area $= \det \begin{pmatrix} 2 & 1 \\ 1 & 2 \end{pmatrix} = 3$.

8. (i)  $\alpha \cdot \beta = -6 + 2 = -4$, $\quad \|\alpha\| = \sqrt{5}$, $\quad \|\beta\| = \sqrt{13}$.

   (ii)  $\cos\Psi = \dfrac{-4}{\sqrt{65}}$.

   (iii)  The perpendicular projection of $\alpha$ on $\beta$ is $\left(\dfrac{12}{13}, \dfrac{-8}{13}\right)$.

   (iv)  Area $= |4 - (-3)| = 7$.

**EXERCISES 1.2** *pages 13–14*

1. $(a + b) \odot \alpha = (a + b) (x, y, z) = (ax + bx, ay + by, az + bz)$
$= (ax, ay, az) \oplus (bx, by, bz) = a \odot \alpha \oplus b \odot \alpha.$

2. Clearly $(0, 0, 0)$ is a zero vector. Conversely, if $(x, y, z)$ is a zero vector, then

$$(a, b, c) \oplus (x, y, z) = (a, b, c);$$
$$a + x = a, \qquad b + y = b, \qquad c + z = c;$$

so

$$(x, y, z) = (0, 0, 0).$$

3. $\alpha = 1\alpha = (1 + 0)\alpha = 1\alpha + 0\alpha = \alpha + 0\alpha.$
Add $-\alpha$ to each side and regroup terms to obtain $\theta = 0\alpha.$

$$\theta = 0\alpha = (1 + (-1))\alpha = 1\alpha + (-1)\alpha = \alpha + (-1)\alpha.$$

Add $-\alpha$ to each side to obtain $-\alpha = (-1)\alpha.$
$$a\theta = a(\alpha + (-\alpha)) = a\alpha + a(-1)\alpha = (a + (-a))\alpha = 0\alpha = \theta.$$

4. (i), (ii), and (iii) are vector spaces but (iv) is not.

5. $\mathscr{P}_2$ and $R_3$ are identical except for notation, where $(a, b, c)$ corresponds to $ax^2 + bx + c.$

6. (i) $A = \{x \in R \mid x = k\pi, k = 0, \pm 1, \pm 2, \ldots \};$
(ii) $B = \{(x, y) \in R \times R \mid (x - 2)^2 + (y - 3)^2 < 1\};$
(iii) $C = \{x, y, z) \in R \times R \times R \mid z > 0\}.$

7. (i) $A$ is the set of all points of $R_3$ *except* the origin;
(ii) $B$ is the set of all points of $R_2$ which lie *below* the line $y = x;$
(iii) $C$ is void.

8. Prove by induction that an $n$-element set has $2^n$ subsets.

9. (i) Only $(2, 9)$ and $(2, 2)$.    (iii) $(4) (5) = 20$ elements.
(ii) $(5) (4) = 20$ elements.    (iv) 4 elements.

10. If $A \subseteq B$, then $B \subseteq A \cup B \subseteq B \cup B = B.$ Conversely, if $A \cup B = B$, then $A \subseteq A \cup B = B.$

11. $(A \cup B) \cap C = (A \cap C) \cup (B \cap C) \subseteq A \cup (B \cap C)$, since $A \cap C \subseteq A.$ If $A \subseteq C$, the middle expression reduces to $A \cup (B \cap C).$

**EXERCISES 1.3** *pages 18–19*

1. (i) $[\alpha, \beta]$ is the plane through $(0, 0, 0)$, $(1, 1, 0)$, $(0, 0, 1)$;

   $[\gamma, \delta]$ is the plane through $(0, 0, 0)$, $(1, 0, 0)$, $(1, 2, 1)$;

   $[\alpha, \beta] \cap [\gamma, \delta]$ is the line through $(0, 0, 0)$ and $(2, 2, 1)$;

   $[\alpha, \beta] \cap [\delta]$ is the origin.

   (ii) $[\alpha, \beta]$ is the plane through $\theta, \alpha, \beta$;

   $[\gamma, \delta]$ is the plane through $\theta, \alpha, \beta$;

   $[\alpha, \beta] \cap [\gamma, \delta]$ is the plane through $\theta, \alpha, \beta$;

   $[\alpha, \beta] \cap [\delta]$ is the line through $\theta, \delta$.

   (iii) $[\alpha, \beta]$ is the plane through $\theta, \alpha, \beta$;

   $[\gamma, \delta]$ is the plane through $\theta, \alpha, \beta$;

   $[\alpha, \beta] \cap [\gamma, \delta]$ is the plane through $\theta, \alpha, \beta$;

   $[\alpha, \beta] \cap [\delta]$ is the line through $\theta, \delta$.

2. (i) $[\alpha, \beta, \gamma] = R_3$.

   (ii) $[\alpha, \beta, \gamma]$ is a plane through $\theta, \alpha, \beta$, since $\gamma = -2\alpha + \beta$.

3. (i), (iii), (v) are subspaces; (ii), (iv), (vi) are not.

4. (i) $\xi = \alpha - 12\beta$,

   (ii) $\xi = -\frac{1}{7}(11\alpha - 18\beta + 4\gamma)$,

   (iii) $\xi \notin [\alpha, \beta, \gamma] = [\alpha, \beta]$.

5. $2x_1 - 5x_2 - 7x_3 = 0$.

6. Note that $\gamma = 3\alpha + 2\beta$ and $\delta = -5\alpha - 4\beta$.
Hence, if
$$\xi = c\gamma + d\delta,$$
then
$$\xi = (3c - 5d)\alpha + (2c - 4d)\beta.$$
Conversely, if
$$\eta = c\alpha + d\beta,$$
then
$$\eta = (2a - \tfrac{5}{2}b)\gamma + (a - \tfrac{3}{2}b)\delta.$$

7. A subspace must be closed under the two vector space operations. Conversely, if a nonvoid subset $S$ is closed under the operations, then $\theta = 0\alpha \in S$ and $-\alpha = (-1)\alpha \in S$ for each $\alpha \in S$. The other vector space properties are inherited in $S$ from $R_3$.

8. Many answers are correct; for example, $\{1, x, x^2\}$ and $\{3, 2 + x, 1 + x^2\}$.

9. $\mathscr{S} \cap \mathscr{T}$ is a subspace, but $\mathscr{S} \cup \mathscr{T}$ is usually not a subspace.

10. Assume that $\mathscr{S} \not\subseteq \mathscr{T}$ and $\mathscr{T} \not\subseteq \mathscr{S}$. Let $\alpha \in \mathscr{S}$ such that $\alpha \notin \mathscr{T}$, and let $\beta \in \mathscr{T}$ such that $\beta \notin \mathscr{S}$. Then $\alpha + \beta \notin \mathscr{S}$ (why is this?), and $\alpha + \beta \notin \mathscr{T}$, so $\alpha + \beta \notin \mathscr{S} \cup \mathscr{T}$. Then $\mathscr{S} \cup \mathscr{T}$ is not a subspace.

**EXERCISES 1.4** *pages 25–26*

  1. (i)  $(6, 1, -1)$.

   (ii)  No solution.

  (iii)  $(2 + c, c, \frac{1}{2})$ for any $c$.

  2. Since
$$\alpha_3 = 2\alpha_1 + \alpha_2,$$
$$[\alpha_1, \alpha_2, \alpha_3] = [\alpha_1, \alpha_2] = \{(x_1 + 2x_2, x_1, 2x_1 + 3x_2)\},$$
a plane through $(0, 0, 0)$, $(1, 1, 2)$ and $(2, 0, 3)$. But $(7, -2, 9) \notin [\alpha_1, \alpha_2, \alpha_3]$, since there is no solution to Exercise 1(ii).

  3. The system is inconsistent – no solution.

  4. The system has infinitely many solutions.

  5. Any linear combination of rows can be achieved by a sequence of elementary row operations. Hence if each row of $B$ is a linear combination of rows of $A$, $B$ can be obtained from $A$ by a sequence of elementary row operations. Conversely, if $B$ can be thus obtained, each row of $B$ must be a linear combination of the rows of $A$.

**EXERCISES 1.5** *pages 31–32*

  1. If $\beta = \sum c_i \alpha_i = \sum d_i \alpha_i$, then $\sum (c_i - d_i)\alpha_i = \theta$. By the property of linear independence, $c_i = d_i$ for $i = 1, 2, 3$.

  2. $\epsilon_1 = \beta_1$,

    $\epsilon_2 = -\beta_1 + \beta_2$,

    $\epsilon_3 = \qquad - \beta_2 + \beta_3$.

  3. $\xi = 9\beta_1 - 7\beta_2 + 3\beta_3$,

    $\eta = 4\epsilon_1 - \quad \epsilon_2 + 3\epsilon_3$.

  4. (i)  If $\theta \in S$ then $S$ is linearly dependent since $1\theta = \theta$.

   (ii)  Any linear combination of vectors of $T$ is a linear combination of vectors of $S$.

(iii)  This is logically equivalent to (ii).

5. If $\sum d_j(c_j\alpha_j) = \theta$, then $d_jc_j = 0$ for each $j$, and so $d_j = 0$ for each $j$.

6. (i) $\{\alpha, \beta, \gamma\}$ is a maximal linearly independent set.

(ii) $\{\alpha, \beta\}$ is a maximal linearly independent set.

(iii) $\{\alpha, \beta\}$ is a maximal linearly independent set.

7. $x = 4$. Geometrically, $(0, 5, x)$ is a vertical line that intersects the plane $[(2, -3, 1), (-4, 1, 2)]$ once.

8. $k = 1$ is the only solution.

9. (i) Verify that $\{\gamma_1, \gamma_2, \gamma_3\}$ is linearly independent and that it spans $R_3$.

(ii) $\epsilon_1 = \frac{2}{3}\gamma_1 + \frac{2}{3}\gamma_2 - \gamma_3,$

$\epsilon_2 = -\gamma_1 \qquad + \gamma_3,$

$\epsilon_3 = \frac{2}{3}\gamma_1 - \frac{1}{3}\gamma_2.$

(iii) $(-2, -3, 2) = 3\gamma_1 - 2\gamma_2 - \gamma_3.$

10. A linear combination of the given polynomials is the zero polynomial if and only if

$$a + \qquad b + \qquad c = 0,$$
$$(r + s)a + (r + t)b + (s + t)c = 0,$$
$$rsa + \qquad rtb + \qquad stc = 0.$$

Because $r$, $s$, and $t$ are distinct, it follows that $a = b = c = 0$.

11. (i) $\xi = (1, 4, -1)$,

(ii) $\xi = [-3, 5, -1]$.


**EXERCISES 1.6** *pages 37–39*

1. To prove (iii) you may use the Cauchy-Schwarz inequality, valid for real numbers: $(x_1y_1 + \ldots + x_ny_n)^2 \le (x_1^2 + \ldots + x_n^2)(y_1^2 + \ldots + y_n^2)$.

2. For (iii) write $d(\alpha, \beta) = \|(\alpha - \gamma) + (\gamma - \beta)\|$ and use the triangle inequality for length.

3. Use the definition of scalar product and properties of real numbers.

4. (ii) Let $\eta = \xi$ and apply the hypothesis.

(iii) Use (ii) with $\xi = \xi_1 - \xi_2$.

5. The converse is true. If $(\alpha \cdot \beta)^2 = (\alpha \cdot \alpha)(\beta \cdot \beta)$, it is possible to expand $(\alpha + c\beta) \cdot (\alpha + c\beta)$ to obtain $(\|\alpha\| + c\|\beta\|)^2$. If $\beta \ne \theta$, $c$ can be chosen to make this expression 0.

6. Expand the left hand member. The sum of the squares of two diagonals of a parallelogram equals the sum of the squares of the lengths of the four sides.

7. Write $(\gamma - \alpha) \cdot (\gamma - \alpha) = (\gamma - \beta + \beta - \alpha) \cdot (\gamma - \beta + \beta - \alpha)$, and expand this to obtain $\|\gamma - \alpha\|^2 = \|\beta - \gamma\|^2 + \|\beta - \alpha\|^2 - 2(\beta - \gamma) \cdot (\beta - \alpha)$. This is the Pythagorean theorem in $E_3$.

8. $(\alpha + \beta) \cdot (\alpha - \beta) = 0$ if and only if $\|\alpha\| = \|\beta\|$.

9. (i) None of the given forms.

 (ii) Right triangle.

 (iii) Isosceles right triangle.

10. (i) $x + 2y + 3z = 0$, a plane orthogonal to $\gamma$.

 (iv) $\xi_1$ and $\xi_2$ span the plane orthogonal to $\gamma$ and through $\theta$.

11. (ii) $(\alpha \cdot \gamma_1) \gamma_1$ is a vector along $\gamma$ of length $|\alpha \cdot \gamma_1| = \|\alpha\| |\cos \Psi|$, and hence it is the orthogonal projection of $\alpha$ on $\gamma$.

12. Let $\gamma = (x, y, z)$ be orthogonal to $\alpha$ and $\beta$.
Then
$$\alpha \cdot \gamma = -x + 2y + 5z = 0,$$
$$\beta \cdot \gamma = 3x - y + z = 0.$$
One solution is $\gamma = (7, 16, -5)$. $\{\alpha, \beta, \gamma\}$ is an orthogonal basis inasmuch as $\alpha \cdot \beta = 0 = \alpha \cdot \gamma = \beta \cdot \gamma$.

13. $\{\alpha, \beta, \gamma\}$ is a basis but not an orthogonal basis since $\alpha \cdot \beta \neq 0$.

14. Many choices for $\beta$ and $\gamma$ are possible. For example, let $\beta = (1, 0, 1)$ and note that $\alpha \cdot \beta = 0$. Let $\xi = (1, 1, -1) \notin [\alpha, \beta]$, and note that $\beta \cdot \xi = 0$. Let
$$\gamma = \xi - \frac{\alpha \cdot \xi}{\alpha \cdot \alpha} \alpha - \frac{\beta \cdot \xi}{\beta \cdot \beta} \beta = \frac{1}{9}(-1, 4, 1).$$
Finally, normalize $\{\alpha, \beta, \gamma\}$ to obtain
$$\alpha_1 = \frac{1}{3}(2, 1, -2),$$
$$\beta_1 = \frac{1}{\sqrt{2}}(1, 0, 1),$$
$$\gamma_1 = \frac{1}{3\sqrt{2}}(-1, 4, 1).$$

15. (i) Let $\xi = \sum x_i \alpha_i$ and $\eta = \sum y_i \alpha_i$ and compute $\xi \cdot \eta$.

 (ii) $\|\xi\|^2 = \sum (\xi \cdot \alpha_i)^2$ expresses the length of any vector in terms of the inner product of that vector with vectors of a normal orthogonal basis.

**EXERCISES 1.7** *page 44*

1. (i) $\dfrac{x-3}{2}=\dfrac{y}{4}=\dfrac{z+1}{5}$.

   (ii) $\cos\Psi_1=\dfrac{2\sqrt{5}}{15}$, $\cos\Psi_2=\dfrac{4\sqrt{5}}{15}$, $\cos\Psi_3=\dfrac{-\sqrt{5}}{3}$.

2. (i) $(x,y,z)=(0,-2,1)+t(-4,3,-2)$.

   (ii) $x=\quad -4t,$

   $y=-2+3t,$

   $z=\quad 1-2t.$

   (iii) $\dfrac{x}{-4}=\dfrac{y+2}{3}=\dfrac{z-1}{-2}.$

3. (i) Direction cosines: $\dfrac{-1}{\sqrt{26}},\dfrac{4}{\sqrt{26}},\dfrac{3}{\sqrt{26}}.$

   (ii) A vector on $L$ can be written

   $\xi=\alpha+t(\beta-\alpha)$ \qquad or \quad $\xi=\beta+s(\alpha-\beta);$

   | | |
   |---|---|
   | $x=-2-\quad t,$ | $x=-3+\quad s,$ |
   | $y=\quad 1+4t,$ | $y=\quad 5-4s,$ |
   | $z=\quad 1+3t;$ | $z=\quad 4-3s;$ |

   $\dfrac{x+2}{-1}=\dfrac{y-1}{4}=\dfrac{z-1}{3}$ \quad or \quad $\dfrac{x+3}{1}=\dfrac{y-5}{-4}=\dfrac{z-4}{-3}.$

   (iii) $(0,-7,-5),\ (-\tfrac{7}{4},0,\tfrac{1}{4}),\ (-\tfrac{5}{3},-\tfrac{1}{3},0).$

   (iv) Determine the vector orthogonal to $L$ and on $L$; it is $-\tfrac{1}{26}(43,10,1)$.

4. Expand $(\alpha+\beta)\cdot(\alpha+\beta)$; this result is the Pythagorean theorem.

5. See Exercise 8 of Section 1.6.

6. $(\alpha+t\beta)\cdot(\alpha+t\beta)=\alpha\cdot\alpha+2t\alpha\cdot\beta+t^2\beta\cdot\beta$. If $\alpha\cdot\beta=0$ then $\|\alpha+t\beta\|$ $\geq\|\alpha\|$. Conversely, if the latter holds for all $t$, then the quadratic expression $2t\alpha\cdot\beta+t^2\beta\cdot\beta$ is nonnegative for all $t$, which implies that $\alpha\cdot\beta=0$.

7. (ii) Let $P,Q\in\bigcap_{C\in\mathscr{F}}C$ where $\mathscr{F}$ is a family of convex sets. For each $C\in\mathscr{F}$ the points $P$, $Q$ and the line segment $PQ$ are in $C$.

**EXERCISES 1.8** *pages 48–49*

1. (i) Respectively, the cosines of the desired angles are $\dfrac{-5}{\sqrt{66}},\dfrac{5}{\sqrt{66}},\dfrac{4}{\sqrt{66}}.$

(ii)  The largest angle is $\dfrac{\pi}{2}$.

(iii)  $(x, y, z) = (4 - 5t, 5 + 5t, 1 + 4t);\ \dfrac{x - 4}{-5} = \dfrac{y - 5}{5} = \dfrac{z - 1}{4}.$

(iv)  $4x - 3y = 1.$

(v)  $2x - 11y + 6z = -41$; unit normal $\dfrac{1}{\sqrt{161}}\,(2, -11, 6).$

(vi)  $\frac{1}{5}$

(vii)  The projection of $\alpha$ onto the normal to the given plane has length 7; use the result of (vi) to obtain $\frac{36}{5}$ as the desired distance.

2.  The Gram-Schmidt process yields $\gamma_1 = (-2, 2, 3)$ and $\gamma_2 = (3, 3, 0)$.

3.  (i)  $8x - 4y - z = 8$;

(ii)  $\pi x + y + z = 2\pi.$

4.  Any point on the line segment from $X_1$ to $X_2$ is of the form $tX_1 + (1 - t)X_2$, $0 \le t \le 1$. If $ax_1 + by_1 + cz_1 \ge d$ and $ax_2 + by_2 + cz_2 \ge d$, then

$$a[tx_1 + (1 - t)x_2] + b[ty_1 + (1 - t)y_2] + c[tz_1 + (1 - t)z_2] \ge td + (1 - t)d = d.$$

5.  (i)  Any line $L$ in $E_3$ has a vector equation $\xi = \alpha + t(\beta - \alpha)$, where $\alpha, \beta \in L$. Let $\xi = (x, y, z)$ and $\rho = (p, q, r)$. Then if $\xi \in L$,

$$f(x, y, z) = \rho \cdot \xi = \rho \cdot \alpha + t\rho \cdot (\beta - \alpha) = m + nt \text{ for suitable } m, n \in R;$$

this is a monotone function of $t$.

(ii)  Use the fact that a linear function is monotone along every line.

6.  $C$ is a tetrahedron with vertices at $O(0, 0, 0)$, $A(6, 0, 0)$, $B(0, 4, 0)$, and $C(0, 0, 3)$. The maximum of $f$ is 6 at $A$ and $C$ and at all points on the edge $AC$. The minimum of $f$ is $-4$ at $B$.

**EXERCISES 1.9** *page 53*

1.

| $\times$ | $\bar{\imath}$ | $\bar{\jmath}$ | $\bar{k}$ |
|---|---|---|---|
| $\bar{\imath}$ | $\theta$ | $\bar{k}$ | $-\bar{\jmath}$ |
| $\bar{\jmath}$ | $-\bar{k}$ | $\theta$ | $\bar{\imath}$ |
| $\bar{k}$ | $\bar{\jmath}$ | $-\bar{\imath}$ | $\theta$ |

2.  If $\beta \times \gamma = \theta$, the result is trivial. Otherwise, $\alpha \times (\beta \times \gamma)$ is perpendicular to $\alpha$ and to $\beta \times \gamma$, and any vector perpendicular to $\beta \times \gamma$ in $E_3$ lies in $[\beta, \gamma]$. Similarly, $(\alpha \times \beta) \times \gamma \in [\alpha, \beta]$ and is perpendicular to $\gamma$.

3. $(-1, 5, -7)$, $(1, -5, 7)$, $(-3, 15, -21)$, $(-36, -45, -27)$, $(19, -20, -17)$.

4. If $\alpha \cdot (\beta \times \gamma) = 0$, then $\alpha$ is perpendicular to $\beta \times \gamma$, so $\alpha \in [\beta, \gamma]$. Conversely if $a\alpha + b\beta + c\gamma = \theta$ with $a \neq 0$, the argument can be reversed. If $a = 0$, $\beta \times \gamma = \theta$, so $\alpha \cdot (\beta \times \gamma) = 0$.

5. Volume $= 0$, so the vectors form a linearly dependent set.

6. A normal vector is $(\alpha - \gamma) \times (\beta - \gamma) = (-3, 15, -21)$. An equation of the plane is $-3(x - 3) + 15(y - 2) - 21(z - 1) = 0$, or $x - 5y + 7z = 0$.

7. Recall that $\beta \times \beta = \theta$. For each $b$ we obtain a parallelogram with base $\|\beta\|$ and height $\|\alpha\| \, |\sin \Psi(\alpha, \beta)|$, so the area is independent of $b$.

8. This involves direct algebraic computation, which is long only for the fifth property.

9. Note first that $\alpha \cdot (\alpha \times (\beta \times \gamma)) = 0 = b\alpha \cdot \beta + c\alpha \cdot \gamma$. Consider the cases $\alpha \cdot \beta = 0$ and $\alpha \cdot \beta \neq 0$.

10. Apply the given result to each of the three terms in the Jacobi identity.

**EXERCISES 1.10** *pages 57–58*

1. For (1) let $c = d = 1$; then let $c = 1$, $d = 0$. For (2) let $c = 0 = d$. For (3) use linearity and (2). For (4) verify linearity for each of the three given combinations.

2. Only ($h$) and ($j$) fail to be linear.

3. For (5) show that $\mathbf{T}(R_3)$ is closed under the vector space operations. For (6) show that the given set is closed under the vector space operations.

4. A plane through the origin; a line through the origin.

5. $(6, 4, 8)$; $(8, 9, 10)$; $(14, 13, 27)$. They are not linearly independent.

6. The results are described in the text presentation of Example (d).

7. Exercises 4 and 5 furnish suitable examples.

8. D can be regarded as the derivative operator on the space $\mathcal{P}_2$ of real polynomials of degree not exceeding 2. Your proofs, however, should not rely on this interpretation.

9. (i) $\mathbf{T}_1$ rotates each point counterclockwise about the origin through the angle $\Psi$;

$\mathbf{T}_2$ projects each point vertically onto the $x$-axis;

$T_3$ projects each point horizontally onto the $y$-axis;

$T_4$ reflects each point across the line $y = x$;

$T_5$ maps each point into itself;

$T_6$ maps each point into $\theta$.

(vii)  Observe that $\{\alpha, \beta\}$ is a basis if and only if $ad \neq bc$.

10. Show that the set of all linear transformations on $R_3$ is closed under the operations of sum and scalar multiple and that the vector space axioms are satisfied.

**EXERCISES 1.11** *pages 64–65*

1. $3D - 2L = \begin{pmatrix} 6 & 0 & 0 \\ -2 & 9 & 0 \\ -2 & -2 & 12 \end{pmatrix}$, $LD = \begin{pmatrix} 0 & 0 & 0 \\ 2 & 0 & 0 \\ 2 & 3 & 0 \end{pmatrix}$, $DL = \begin{pmatrix} 0 & 0 & 0 \\ 3 & 0 & 0 \\ 4 & 4 & 0 \end{pmatrix}$,

2. $LY = \begin{pmatrix} 0 \\ 0 \\ 1 \end{pmatrix}$, $AE = \begin{pmatrix} \sum a_{1j} \\ \sum a_{2j} \\ \sum a_{3j} \end{pmatrix}$.

3. $AZ = Z = ZA$.

4. $AI = A = AI$.

5. $D^2 = \begin{pmatrix} 4 & 0 & 0 \\ 0 & 9 & 0 \\ 0 & 0 & 16 \end{pmatrix}$, $D^3 = \begin{pmatrix} 8 & 0 & 0 \\ 0 & 27 & 0 \\ 0 & 0 & 64 \end{pmatrix}$.

6. $DA = \begin{pmatrix} 2a_{11} & 2a_{12} & 2a_{13} \\ 3a_{21} & 3a_{22} & 3a_{23} \\ 4a_{31} & 4a_{32} & 4a_{33} \end{pmatrix}$, $AD = \begin{pmatrix} 2a_{11} & 3a_{12} & 4a_{13} \\ 2a_{21} & 3a_{22} & 4a_{23} \\ 2a_{31} & 3a_{32} & 4a_{33} \end{pmatrix}$.

7. $L^2 = \begin{pmatrix} 0 & 0 & 0 \\ 0 & 0 & 0 \\ 1 & 0 & 0 \end{pmatrix}$, $L^3 = \begin{pmatrix} 0 & 0 & 0 \\ 0 & 0 & 0 \\ 0 & 0 & 0 \end{pmatrix}$.

8. $U_{21}A = \begin{pmatrix} 0 & 0 & 0 \\ a_{11} & a_{12} & a_{13} \\ 0 & 0 & 0 \end{pmatrix}$, $AU_{21} = \begin{pmatrix} a_{12} & 0 & 0 \\ a_{22} & 0 & 0 \\ a_{32} & 0 & 0 \end{pmatrix}$.

Row 2 of the product $U_{21}A$ is row 1 of $A$; column 1 of $AU_{21}$ is column 2 of $A$.

9. $LU_{21} = \begin{pmatrix} 0 & 0 & 0 \\ 0 & 0 & 0 \\ 1 & 0 & 0 \end{pmatrix}.$

10. $X = \begin{pmatrix} 1 \\ 1 \\ c \end{pmatrix}$, $c$ arbitrary.

11. First observe that $\mathcal{M}_{3 \times 3}$ is closed under sum and scalar multiple.

12. Let $A = (a_{ij})$. Then $A = \sum a_{ij} U_{ij}$, so the matrices $U_{ij}$ span $\mathcal{M}_{3 \times 3}$. Linear independence follows easily from the definition of equality of matrices.

**EXERCISES 1.12** *pages 69–70*

1. (i) $\begin{pmatrix} 3 & -1 & -\frac{1}{2} \\ -1 & 1 & 0 \\ -\frac{1}{2} & 0 & -1 \end{pmatrix}$,   (ii) $\begin{pmatrix} 0 & \frac{1}{2} & 1 \\ \frac{1}{2} & 0 & \frac{1}{2} \\ 1 & \frac{1}{2} & 0 \end{pmatrix}.$

2. Compute $X^t A X$ by matrix multiplication.

3. (iv) $(AB)^t = (c_{ij})$ where $c_{ij} = \sum a_{jk} b_{ki}$, and
    $B^t A^t = (d_{ij})$ where $d_{ij} = \sum b_{ki} a_{jk} = c_{ij}.$

4. $(A + A^t)^t = A^t + (A^t)^t = A + A^t$, symmetric;
    $(AA^t)^t = (A^t)^t A^t = AA^t$, symmetric.

5. (i) $(A + B)^t = A^t + B^t = A + B.$
    (ii) $(AB)^t = B^t A^t = BA$, so $AB$ is symmetric if and only if $AB = BA.$

6. (i) Analogous to 5(i).
    (ii) Write $A = \frac{1}{2}(A + A^t) + \frac{1}{2}(A - A^t).$
    (iii) $(A^2)^t = (A^t)^2 = (\pm A)^2 = A^2$, symmetric.

**EXERCISES 1.13** *pages 80–81*

1. $R$ consists of all points in the first quadrant except those below the line segments joining $(0, 12)$ to $(2, 8)$ to $\frac{1}{11}(126, 23)$ to $(24, 0)$.

2. (i) $(2, 8)$,    (iii) $\frac{1}{11}(126, 23)$,
    (ii) $(0, 12)$,    (iv) no minimum exists.

3. $R$ consists of all points of the first quadrant on or below the line segments joining $(0, 360)$ to $(120, 300)$ to $\frac{900}{7}(2, 1)$ to $(300, 0)$.

(i)  (120, 300),

(ii)  (120, 300),

(iii)  $\frac{900}{7}(2, 1)$.

4. (ii)  $A + A^2 + A^3 = \begin{pmatrix} 2 & 4 & 1 & 3 & 3 \\ 2 & 3 & 2 & 2 & 4 \\ 3 & 4 & 2 & 3 & 5 \\ 2 & 2 & 2 & 1 & 4 \\ 1 & 2 & 1 & 1 & 2 \end{pmatrix}$.

5. $B^2 = \begin{pmatrix} 0 & 0 & 2 & 0 & 0 & 0 \\ 1 & 0 & 0 & 0 & 1 & 0 \\ 1 & 1 & 0 & 2 & 0 & 1 \\ 1 & 0 & 0 & 0 & 1 & 0 \\ 1 & 2 & 1 & 1 & 0 & 0 \\ 0 & 1 & 1 & 1 & 0 & 0 \end{pmatrix}$.

6. A zero row sum means "no influence." A zero column sum means "influenced by nobody."

7. (i)  Let $A$ and $B$ be Markov matrices, and let $AB = (c_{ij})$, where $c_{ij} = \sum_k a_{ik}b_{kj}$. Then

$$\sum_j c_{ij} = \sum_j \sum_k a_{ik}b_{kj} = \sum_k \left( a_{ik} \left( \sum_j b_{kj} \right) \right) = \sum_k a_{ik} = 1.$$

Also $0 \le c_{ij} \le 1$.

(ii)  $\sum (aa_{ij} + bb_{ij}) = a \sum a_{ij} + b \sum b_{ij} = a + b = 1$.
Also $0 \le (aa_{ij} + bb_{ij}) \le 1$.

8. Follow the text example, supplying details of computation.

# 2

**EXERCISES 2.1** *pages 91–92*

1. Show that $\mathscr{S}$ satisfies Definition 2.1. For $\mathscr{K}$, use Theorem 2.2.

2. Verify that if $y_1$ and $y_2$ are solutions, so is $y_1 + y_2$ and $cy_1$ for any $c$. The other properties of a vector space follow routinely.

3. Let $a_0x^k + a_1x^{k-1} + \ldots + a_k$ correspond to $(a_0, a_1, \ldots, a_k)$, and observe that this correspondence preserves the vector space operations.

4. Consider $(\alpha'' + \alpha) + \alpha'$, where $\alpha'$ and $\alpha''$ satisfy Property (4).

5. Begin by considering $k(\alpha + (-\alpha))$.

6. All are subspaces except (i) and (iii).

7. Only (iii) and (iv) are subspaces.

8. Only (i) and (iv) are subspaces.

9. Only (i) is a subspace.

10. (i) $\mathscr{S} \cap \mathscr{T}$ is the line through $(0, 0, 0)$ and $(2, 2, 1)$; $\mathscr{S} + \mathscr{T} = R_3$.

   (ii) $\mathscr{S} \cap \mathscr{T} = \mathscr{S} = \mathscr{T} = \mathscr{S} + \mathscr{T}$, a plane.

   (iii) $\mathscr{S} \cap \mathscr{T} = \mathscr{S} = \mathscr{T} = \mathscr{S} + \mathscr{T}$, a plane.

11. (i) If $\xi \in (\mathscr{S} \cap \mathscr{U}) + (\mathscr{T} \cap \mathscr{U})$, then $\xi = \sigma + \tau$ where $\sigma \in \mathscr{S} \cap \mathscr{U}$, $\tau \in \mathscr{T} \cap \mathscr{U}$.

   (ii) Consider three distinct lines through $(0, 0)$ in $R_2$.

   (iii) Let $\mathscr{S} \subseteq \mathscr{U}$, and let $\eta \in (\mathscr{S} + \mathscr{T}) \cap \mathscr{U}$. Then $\eta = \sigma + \tau$ where $\sigma \in \mathscr{S}$, $\tau \in \mathscr{T}$, $\eta \in \mathscr{U}$. Show that $\tau \in \mathscr{U}$ and then complete the proof.

12. Yes.


## EXERCISES 2.2 *pages 95–96*

1. (i) Exclude one of the first three vectors.

   (ii) All four vectors.

   (iii) Select any two vectors.

2. Various choices are possible; for example, adjoin $(0, 0, 0, 1)$ in (i), and adjoin $(0, 1, 0, 0)$ and $(0, 0, 1, 0)$ in (ii).

3. Only (ii) is linearly dependent.

4. If $\sum c_i \alpha_i = \theta$, then $c_1 = 0$, $c_1 + c_2 = 0$, . . . , $c_1 + \ldots + c_n = 0$. Each $c_i = 0$.

5. Recall that a polynomial of degree $k$ has at most $k$ zeros.

6. Suppose that $c_1 + c_2 \sqrt{2} = 0$. If $c_1$ and $c_2$ are rational, both must be 0. But if $c_1$ and $c_2$ can be irrational, let $c_1 = \sqrt{2}$, $c_2 = -1$.

7. Let $\mathscr{S} = [S]$. If $S$ is linearly dependent, apply Theorem 2.8 to discard an appropriate vector from $S$. This process can be repeated, if necessary, and eventually it terminates.

8. If $\{f, f'\}$ is linearly dependent, $f' = kf$ for some scalar $k$. Hence the solution consists of all functions of the form $f(x) = ae^{kx}$.

9. Any linear combination of vectors in $S \cup T$ is of the form $\sigma + \tau$, where $\sigma \in \mathcal{S}$ and $\tau \in \mathcal{T}$. If $\sigma + \tau = \theta$, then $\sigma = \theta = \tau$ since $\mathcal{S} \cap \mathcal{T} = [\theta]$.

10. Write the $m$ vectors as rows of an $m$-by-$n$ matrix $A$. Since $m > n$ any echelon form of $A$ must have zero rows in at least the last $m - n$ rows, so the given vectors are linearly dependent.

## EXERCISES 2.3 *pages 102–104*

1. Reduce the matrix of row vectors to echelon form: dimension $= 3$.

2. Respectively, the dimensions are $n - 1$, $n - 1$, and $n - 2$.

3. (i) See Exercise 5 of Section 1.5.

   (ii) Assume $\sum c_i \beta_i = \theta$. Use the linear independence of $\{\alpha_1, \ldots, \alpha_n\}$ to show that each $c_i = 0$.

4. (i) If $\sum c_i(\alpha_i - \beta) = \theta$, then $\sum c_i = 0$ since $\beta \notin \mathcal{S}$. Then each $c_i = 0$, since $\{\alpha_1, \ldots, \alpha_k\}$ is linearly independent.

   (ii) $\dim(\mathcal{S} + \mathcal{T}) = k + 1$ and $\dim(\mathcal{S} \cap \mathcal{T}) = k - 1$.

5. A basis for $\mathcal{T}$ is a basis for $\mathcal{S}$.

6. $\begin{aligned} \epsilon_1 &= \alpha_1 - \alpha_2 \\ \epsilon_2 &= \alpha_2 - \alpha_3 \\ \epsilon_3 &= \alpha_3 - \alpha_4 \\ \epsilon_4 &= \alpha_4 \end{aligned}$ $\qquad \begin{aligned} \alpha_1 &= \epsilon_1 + \epsilon_2 + \epsilon_3 + \epsilon_4 \\ \alpha_2 &= \epsilon_2 + \epsilon_3 + \epsilon_4 \\ \alpha_3 &= \epsilon_3 + \epsilon_4 \\ \alpha_4 &= \epsilon_4 \end{aligned}$

7. If $A$ is linearly independent, Exercise 5 shows that $[A] = \mathcal{V}_n$. Conversely, if $[A] = \mathcal{V}_n$, a linearly independent subset $S$ of $A$ spans $\mathcal{V}_n$, and because of the dimensions involved, $S = A$.

8. If $[A] \neq \mathcal{V}_n$ adjoin to $A$ any vector $\beta_1 \in \mathcal{V}_n$ such that $\beta_1 \notin [A]$. Then $A_1 = \{\beta_1\} \cup A$ is linearly independent. Extend this construction until a basis for $\mathcal{V}_n$ is obtained.

9. A basis for $\mathcal{V}_n$ must span $\mathcal{V}_n$ and be linearly independent. The latter condition implies that a basis is a minimal spanning set. The former condition implies that a basis is a maximal linearly independent set.

10. (i) $\mathscr{S} + \mathscr{T} = \mathscr{V}_n$; linear independence of $\{\alpha_1, \ldots, \alpha_n\}$ implies that $\mathscr{S} \cap \mathscr{T} = [\theta]$.

(ii) $\{\alpha_1, \ldots, \alpha_k, \beta_1, \ldots, \beta_m\}$ spans $\mathscr{S} + \mathscr{T} = \mathscr{V}_n$. The condition $\mathscr{S} \cap \mathscr{T} = [\theta]$ implies that this set is linearly independent.

11. The set union of a basis for $\mathscr{S}$ and a basis for $\mathscr{T}$ is a basis for $\mathscr{V}$.

12. (i) Vectors are elements of $F_2$; scalars are elements of $F_1$. The vector space axioms are easily verified.

(ii) $Q$ is of dimension 1 over $Q$. $C$ is of dimension 2 over $R$ since $\{1, i\}$ is a basis for $C$ over $R$. $C$ is of dimension 1 over $C$. But $R$ is of infinite-dimension over $Q$.

13. $f\left(\sum a_i x^{k-1}\right) = (a_0, a_1, \ldots, a_k)$.   $g(a + ib) = (a, b)$.

The fact that these are isomorphisms should be verified in detail.

14. (i) If $\eta = \sum b_i \beta_i$ then $\eta = \mathbf{T}\left(\sum b_i \alpha_i\right)$.

(ii) If $\mathbf{T}(\xi) = \sum b_i \beta_i = \mathbf{T}(\eta)$ then $\mathbf{T}(\xi - \eta) = \theta$. But the definition of $\mathbf{T}$ guarantees that $\theta$ is the only vector that $\mathbf{T}$ maps into $\theta$.

15. The given condition is equivalent to the condition that $h$ is one-to-one.

16. Let $S$ be linearly independent in $\mathscr{V}$, and suppose that $\sum c_i f(\sigma_i) = \theta$ in $\mathscr{W}$, where $\sigma_i \in S$. Then $f\left(\sum c_i \sigma_i\right) = \theta$, so $\sum c_i \sigma_i = \theta$, and each $c_i = 0$.

**EXERCISES 2.4** *pages 108–109*

1. (ii) $f_k(\alpha, \alpha) > 0$ for $\alpha \neq \theta$ if and only if $k > 1$.

2. For Example (c), observe that $x_1^2 - 4x_1 x_2 + 8x_2^2 = (x_1 - 2x_2)^2 + 4x_2^2$.

3. The inequalities stated are obtained directly from the Schwarz inequality for the two inner products of Examples (a) and (b).

4. $\langle \alpha - \beta, \xi \rangle = 0$ for all $\xi$, and hence for $\xi = \alpha - \beta$.

6. Equality holds if and only if $\dfrac{|\langle \alpha, \beta \rangle|}{\|\alpha\| \|\beta\|} = 1$. But then $\Psi(\alpha, \beta) = 0$ or $\pi$, so $\alpha$ and $\beta$ are collinear.

7. Consider $\langle c\alpha, c\alpha \rangle$ for (a); (b) follows directly from Property (c) of an inner product.

9. Each property of distance follows from the corresponding property of length.

10. Expand the two inner products on the left. See Exercise 6 of Section 1.6.

11. Let the consecutive vertices of the quadrilateral determine vectors $\theta$, $\alpha$, $\beta$, $\gamma$. The theorem then becomes the trivial statement:

$$\left\|\frac{\beta}{2}\right\| = \left\|\frac{\alpha + \beta}{2} - \frac{\alpha}{2}\right\| \text{ and } \left\|\frac{\alpha}{2} - \frac{\gamma}{2}\right\| = \left\|\frac{\alpha + \beta}{2} - \frac{\beta + \gamma}{2}\right\|.$$

12. (ii)  Observe that

$$\max |aa_i| = |a|^{\cdot}\max |a_i|,$$

and

$$\max |a_i + b_i| \leq \max |a_i| + \max |b_i|.$$

13. Observe that

$$\max |af(x)| = |a| \max |f(x)|,$$

and

$$\max |f(x) + g(x)| \leq \max |f(x)| + \max |g(x)|.$$

**EXERCISES 2.5** *pages 113–114*

1. Relative to a normal orthogonal basis the formula becomes

$$\langle \xi, \eta \rangle = \|\xi\| \|\eta\| \cos \Psi.$$

Then

$$\|\xi - \eta\|^2 = \langle \xi - \eta, \xi - \eta \rangle = \|\xi\|^2 + \|\eta\|^2 - 2 \langle \xi, \eta \rangle.$$

2. To prove (c) suppose that $\sum c_i \sigma_i = \theta$, where the $\sigma_i$ are mutually orthogonal. Then for each $j$, $\left\langle \sum c_i \sigma_i, \sigma_j \right\rangle = 0 = c_j.$

3. The diagonals of a parallelogram are perpendicular if and only if the parallelogram is equilateral. See Exercise 8 of Section 1.6.

4. $\langle \xi, \eta \rangle = 0$ if and only if $\|\xi + \eta\|^2 = \|\xi\|^2 + \|\eta\|^2.$

5. See Exercise 6 of Section 1.7.

6. Many constructions are possible. For example, choose $\alpha_3 = (0, 0, 0, 1)$.

Let $\xi = (1, 0, 0, 0)$, and let $\alpha_4 = \xi - \sum \frac{\langle \alpha_i, \xi \rangle}{\langle \alpha_i, \alpha_i \rangle} \alpha_i$. Then normalize.

7. (i)  Let $\xi = \sum x_j \alpha_j$. For each $i$, $\langle \xi, \alpha_i \rangle^2 = x_i^2$, and $\|\xi\|^2 = x_1^2 + \ldots + x_n^2$ since the inner product assumes standard form.

   (ii)  $\langle \xi, \eta \rangle = \sum x_k y_k$ where $x_k = \langle \xi, \alpha_k \rangle$ and $y_k = \langle \alpha_k, \eta \rangle.$

8. (iii)  One decomposition $\xi = \sigma + \tau$ is assured by the Gram-Schmidt process. If $\xi = \sigma_1 + \tau_1$, then $\sigma - \sigma_1 = \tau_1 - \tau \in \mathscr{S}$. But $\tau_1 - \tau$ is orthogonal to $\mathscr{S}$, so $\tau = \tau_1$ and $\sigma = \sigma_1$.

9. (ii)  Use Exercise 8.

(iii)  Show that $\mathscr{S} \subseteq (\mathscr{S}^\perp)^\perp$. If $\mathscr{V}$ is finite dimensional, $\mathscr{V} = \mathscr{S} \oplus \mathscr{S}^\perp$, so also $\mathscr{V} = \mathscr{S}^\perp \oplus (\mathscr{S}^\perp)^\perp$. Hence dim $\mathscr{S} = \dim(\mathscr{S}^\perp)^\perp$, and equality holds.

(iv)  If $\xi \in (\mathscr{S} + \mathscr{T})^\perp$, $\langle \xi, \sigma \rangle = 0 = \langle \xi, \tau \rangle$, so $\xi \in \mathscr{S}^\perp \cap \mathscr{T}^\perp$. The reverse argument also works without difficulty.

(v)  Show that $\mathscr{S}^\perp + \mathscr{T}^\perp \subseteq (\mathscr{S} \cap \mathscr{T})^\perp$. Then use dim $\mathscr{S}^\perp = n - \dim \mathscr{S}$, Theorem 2.13, and (iv) to prove equality.

10. (ii)  Relative to the orthogonal basis $N$, $p$ has coordinates $\left( \dfrac{1}{6\sqrt{5}}, \dfrac{1}{2\sqrt{3}}, \dfrac{1}{3} \right)$

and $q$ has coordinates $\left( -\dfrac{1}{\sqrt{5}}, -\dfrac{2}{\sqrt{3}}, -1 \right)$. Then $\langle p, q \rangle = \displaystyle\int_0^1 x^2(2x - 6x^2)\,dx$

$$= \frac{-7}{10} = \left( \frac{1}{6\sqrt{5}}, \frac{1}{2\sqrt{3}}, \frac{1}{3} \right) \cdot \left( -\frac{1}{\sqrt{5}}, -\frac{2}{\sqrt{3}}, -1 \right).$$

# 3

**EXERCISES 3.1** *pages 121–122*

1. $T(\theta) = T(0\alpha) = 0T(\alpha) = \theta$.

2. $(TZ)(\xi) = \theta = (ZT)\,(\xi) = Z(\xi)$ for every $\xi$.

3. (i)  $J$ is linear, but it maps $\mathscr{P}_2$ into $\mathscr{P}_3$.

(ii)  $D$ is a linear operator on $\mathscr{P}$, and if $p$ is a polynomial of degree $k$, then $D^{k+1}p = \theta$. But there is no fixed $n$ such that $D^n = Z$ since $\mathscr{P}$ contains polynomials of all degrees; so $D$ is not nilpotent on $\mathscr{P}$.

4. Geometrically, $T_\omega T_\Psi$ is a counterclockwise rotation of the plane around $\theta$ through the angle $\Psi$, followed by a rotation through the angle $\omega$. The algebraic derivation of this fact uses the trigonometric sum identities.

5. $T(x, y) = T(x\epsilon_1 + y\epsilon_2) = xT(\epsilon_1) + yT(\epsilon_2) = (-4x + y, 5x - 2y)$.

6. (i)  $T$ is linear over $R$.

(ii)  $T$ is not linear over $C$.

(iii)  In (ii) $T$ satisfies (2) but not (3).

7. Let $h(x, y) = \dfrac{x^2}{x^2 + y^2}(x, y)$   if $(x, y) \neq (0, 0)$.

$\qquad\qquad\quad = (0, 0)$   \qquad\qquad if $(x, y) = (0, 0)$.

8. **T** is a reflection across the line $y = x$. **S** is a vertical projection onto the $x$ axis, and **R** is a horizontal projection onto the $y$ axis.

10. (ii)  $\mathbf{DM}p(x) = \mathbf{D}(xp(x)) = p(x) + xp'(x),$

$\mathbf{MD}p(x) = xp'(x).$ Hence $(\mathbf{DM} - \mathbf{MD})p(x) = p(x) = \mathbf{I}p(x).$

11. If $c_1\xi + c_2\mathbf{T}\xi + \ldots + c_p\mathbf{T}^{p-1}\xi = \theta$, let $c_j$ be the first nonzero coefficient. Then

$$\mathbf{T}^{p-j}(c_j\mathbf{T}^{j-1}\xi + \ldots + c_p\mathbf{T}^{p-1}\xi) = \mathbf{T}^{p-j}\theta = \theta$$

$$= c_j\mathbf{T}^{p-1}\xi \neq \theta, \text{ a contradiction.}$$

Hence each $c_i = 0$.

12. $f(x_1, \ldots, x_n) = (c_1, \ldots, c_n) \cdot (x_1, \ldots, x_n)$, so this example is the same as Example (g) if the standard basis is used for $E_n$.

13. To prove that $N_1(\mathbf{T}) > 0$ if $\mathbf{T} \neq \mathbf{Z}$, there exists $\alpha$ such that $\mathbf{T}\alpha \neq \theta$. The vector $\alpha_1 = N(\alpha)^{-1}\alpha$ satisfies $N(\alpha_1) = 1$. Then $N_1(\mathbf{T}) = \max N(\mathbf{T}\xi) > 0$ since $N(\mathbf{T}\alpha_1) > 0$. To prove the subadditive norm inequality, we have

$$N_1(\mathbf{S} + \mathbf{T}) = \max N((\mathbf{S} + \mathbf{T})\xi) = \max N(\mathbf{S}\xi + \mathbf{T}\xi) \leq \max (N(\mathbf{S}\xi) + N(\mathbf{T}\xi))$$

$$\leq \max N(\mathbf{S}\xi) + \max N(\mathbf{T}\xi) = N_1(\mathbf{S}) + N_1(\mathbf{T}).$$

## EXERCISES 3.2 *page 127*

1. Example (c) is of rank 2 and nullity 0; $2 + 0 = \dim R_2$.
   Example (i) is of rank $n$ and nullity 1; $n + 1 = \dim \mathscr{P}_n$.
   Example (j) is of rank 3 and nullity 0; $3 + 0 = \dim R_3$.

2. $\mathscr{R}(\mathbf{D}) = \mathscr{P}, \mathscr{N}(\mathbf{D}) = \mathscr{P}_0; \mathscr{R}(\mathbf{M}) = \{p \in \mathscr{P}|p(0) = 0\}, \mathscr{N}(\mathbf{M}) = [\theta].$

3. $\mathbf{T}^2(a_1, a_2, a_3) = (0, 0, 4a_3), \mathbf{T}^3(a_1, a_2, a_3) = (0, 0, 8a_3).$

$$\mathscr{N}(\mathbf{T}) = [\epsilon_1], \mathscr{N}(\mathbf{T}^2) = [\epsilon_1, \epsilon_2] = \mathscr{N}(\mathbf{T}^3).$$

$$\mathscr{R}(\mathbf{T}) = [\epsilon_1, \epsilon_2], \mathscr{R}(\mathbf{T}^2) = [\epsilon_3] = \mathscr{R}(\mathbf{T}^3).$$

$$[\theta] \subset \mathscr{N}(\mathbf{T}) \subset \mathscr{N}(\mathbf{T}^2) = \mathscr{N}(\mathbf{T}^3) = \ldots,$$

$$R_3 \supset \mathscr{R}(\mathbf{T}) \supset \mathscr{R}(\mathbf{T}^2) = \mathscr{R}(\mathbf{T}^3) = \ldots,$$

and

$$R_3 = \mathscr{R}(\mathbf{T}^2) \oplus \mathscr{N}(\mathbf{T}^2).$$

4. If $r(\mathbf{T}) = 1$, then $\mathscr{R}(\mathbf{T}) = [\alpha]$ for some $\alpha \neq \theta$. For any $\xi$

$\mathbf{T}(\xi) = k\alpha$, where $k$ depends on $\xi$,

$\mathbf{T}^2(\xi) = k\mathbf{T}(\alpha) = kc\alpha = (c\mathbf{T})(\xi)$; $c$ depends on $\alpha$.

Hence $\mathbf{T}^2 = c\mathbf{T}$, where $c$ is such that $\mathbf{T}\alpha = c\alpha$ and $[\alpha] = \mathscr{R}(\mathbf{T})$. The converse is not valid.

5. If $\xi \in \mathcal{N}(\mathbf{T})$, $\mathbf{T}\xi = \theta$, so $\mathbf{ST}\xi = \theta$; then $\xi \in \mathcal{N}(\mathbf{ST})$.
If $\eta \in \mathcal{R}(\mathbf{ST})$, $\eta = \mathbf{ST}\xi$ for some $\xi$; let $\zeta = \mathbf{T}\xi$, so that $\eta = \mathbf{S}\zeta \in \mathcal{R}(\mathbf{S})$.

7.  (ii)  From (i): $r(\mathbf{S} + \mathbf{T}) \le \dim(\mathcal{R}(\mathbf{S}) + \mathcal{R}(\mathbf{T}))$

$$= r(\mathbf{S}) + r(\mathbf{T}) - \dim(\mathcal{R}(\mathbf{S}) \cap \mathcal{R}(\mathbf{T})).$$

(iii)  From (ii): $n(\mathbf{S} + \mathbf{T}) = n - r(\mathbf{S} + \mathbf{T}) \ge n - r(\mathbf{S}) - r(\mathbf{T}) = n(\mathbf{S}) + n(\mathbf{T}) - n$.

8. Part (ii) follows from (i) and Theorem 3.4. The second inequality of (i) follows from Exercise 6. To prove the first inequality of (i), choose a basis $\{\alpha_1, \ldots, \alpha_t\}$ for $\mathcal{N}(\mathbf{T})$ and extend it to a basis $\{\alpha_1, \ldots, \alpha_t, \alpha_{t+1}, \ldots, \alpha_s\}$ for $\mathcal{N}(\mathbf{ST})$. Then show that $\{\mathbf{T}\alpha_{t+1}, \ldots, \mathbf{T}\alpha_s\}$ is a basis for the subspace $\mathbf{T}(\mathcal{N}(\mathbf{ST}))$, which is itself a subspace of $\mathcal{N}(\mathbf{S})$.

9. The desired assertions follow easily from the observations that

$$\mathcal{R}(\mathbf{S} + \mathbf{T}) = R_3 = \mathcal{R}(\mathbf{T}) \text{ and } \mathcal{R}(\mathbf{ST}) = [(\epsilon_1, \epsilon_2)] = \mathcal{R}(\mathbf{S}).$$

**EXERCISES 3.3** *pages 132–133*

1. $\mathbf{T}(x, y) = (ax + cy, bx + dy)$. The condition that $\mathbf{T}$ is nonsingular is that there exist no nontrivial solutions to the system

$$ax + cy = 0,$$
$$bx + dy = 0.$$

Geometrically, the condition $ad \ne bc$ means that the image vectors $\mathbf{T}\epsilon_1$ and $\mathbf{T}\epsilon_2$ have different slopes.

2. $(\mathbf{T}^{-1})^{-1} = (\mathbf{T}^{-1})^{-1}(\mathbf{T}^{-1}\mathbf{T}) = ((\mathbf{T}^{-1})^{-1}\mathbf{T}^{-1})\mathbf{T} = \mathbf{IT} = \mathbf{T}$.

3. (i)  If $\mathbf{ST}$ is nonsingular there is a linear mapping $\mathbf{R}$ from $\mathcal{R}(\mathbf{ST})$ to $\mathcal{X}$ such that $\mathbf{R}(\mathbf{ST}) = \mathbf{I}$. Then $(\mathbf{RS})\mathbf{T} = \mathbf{I}$, so $\mathbf{T}$ is nonsingular and $\mathbf{RS} = \mathbf{T}^{-1}$. Hence $\mathbf{T}(\mathbf{RS}) = \mathbf{I} = (\mathbf{TR})\mathbf{S}$, and $\mathbf{S}$ is nonsingular. The converse is easy.

(ii)  $(\mathbf{T}^{-1}\mathbf{S}^{-1})(\mathbf{ST}) = \mathbf{I}$, where $\mathbf{T}^{-1}$ and $\mathbf{S}^{-1}$ exist by (i).

4. A nonsingular mapping $\mathbf{T}$ on a $k$-dimensional space has rank $k$; hence $r(\mathbf{TT}_1) = r(\mathbf{T}_1)$. Also $r(\mathbf{T}_2\mathbf{T}) = r(\mathbf{T}_2)$ since $\mathbf{T}$ is nonsingular from $\mathcal{R}(\mathbf{T}_1)$ onto $\mathcal{R}(\mathbf{T})$ (which are spaces of the same dimension) and since $\mathbf{T}_2$ is a mapping from $\mathcal{R}(\mathbf{T})$.

5. Any linearly independent set in $\mathcal{V}_n$ is part of a basis for $\mathcal{V}_n$, and $\mathbf{T}$ is nonsingular if and only if it carries a basis into a basis, by Theorem 3.7.

6. (i)    $\mathbf{J}$ and $\mathbf{JD}$ have domain $\mathcal{P}$, range $\mathcal{P}_0$,

$\mathbf{D}$, $\mathbf{DJ}$, and $\mathbf{D}_0\mathbf{J}$ have domain $\mathcal{P}$, range $\mathcal{P}$,

$\mathbf{D}_0$ has domain $\mathcal{P}_0$, range $\mathcal{P}$,

$\mathbf{JD}_0$ has domain $\mathcal{P}_0$, range $\mathcal{P}_0$.

(ii)  $\mathbf{I} = \mathbf{DJ},\ \mathbf{D}_0\mathbf{J},\ \mathbf{JD}_0.$

(iii)  All except $\mathbf{D}$ and $\mathbf{JD}$ are nonsingular.

7. If $\alpha = \theta$ or $\beta = \theta$, the result is immediately clear. Since $\mathbf{T}$ is orthogonal, $\langle \alpha, \alpha \rangle = \langle \mathbf{T}\alpha, \mathbf{T}\alpha \rangle = \langle a\alpha, a\alpha \rangle = a^2 \langle \alpha, \alpha \rangle$. So $a = \pm 1$, and similarly $b = \pm 1$. Since $a$ and $b$ are distinct, $ab = -1$. Then $\langle \alpha, \beta \rangle = \langle \mathbf{T}\alpha, \mathbf{T}\beta \rangle = -\langle \alpha, \beta \rangle$, so $\langle \alpha, \beta \rangle = 0$.

8. If $\mathbf{T}^{p-1}\xi \neq \theta$, the space $[\xi, \mathbf{T}\xi, \ldots, \mathbf{T}^{p-1}\xi]$ is $\mathbf{T}$-invariant.

9. Let $\xi \in \mathscr{R}(\mathbf{T}^{p-k})$; then $\xi \in \mathscr{N}(\mathbf{T}^k)$, $k = 0, \ldots, p-1$.

10. If $\eta \in \mathscr{R}(\mathbf{T})$, $\eta = \mathbf{T}\xi$ for some $\xi \in \mathscr{V}$. Then $\mathbf{T}(\eta) = \mathbf{T}^2\xi = \mathbf{T}\xi = \eta$.

11. (ii)  $(\mathbf{E}_1 + \mathbf{E}_2)(\xi) = \mathbf{E}_1\xi + \mathbf{E}_2\xi = \mu_1 + \mu_2 = \xi$, so $\mathbf{E}_1 + \mathbf{E}_2 = \mathbf{I}$.
$(\mathbf{E}_1\mathbf{E}_2)(\xi) = \mathbf{E}_1\mu_2 = \theta = \mathbf{E}_2\mu_1 = \mathbf{E}_2\mathbf{E}_1\xi$, so $\mathbf{E}_1\mathbf{E}_2 = \mathbf{Z} = \mathbf{E}_2\mathbf{E}_1$.

**EXERCISES 3.4** *pages 139–141*

1. (i)  $A = \begin{pmatrix} 1 & 1 \\ -1 & 1 \end{pmatrix}.$

(iii)  $X = \begin{pmatrix} x_1 \\ x_2 \end{pmatrix}, \quad Y = \begin{pmatrix} \dfrac{x+y}{3} \\ \dfrac{2x-y}{3} \end{pmatrix}.$

(ii)  $B = \begin{pmatrix} \dfrac{4}{3} & -\dfrac{2}{3} \\ \dfrac{5}{3} & \dfrac{2}{3} \end{pmatrix}.$

(iv)  $BX = \begin{pmatrix} \dfrac{4x-2y}{3} \\ \dfrac{5x+2y}{3} \end{pmatrix}, \quad AY = \begin{pmatrix} x \\ \dfrac{x-2y}{3} \end{pmatrix}.$

2. (i)  $A = \begin{pmatrix} 2 & 1 & 1 \\ -2 & 1 & 3 \\ 3 & 1 & -1 \end{pmatrix}.$

(iii)  $\begin{aligned} \mathbf{T}\alpha_1 &= 3\alpha_1, \\ \mathbf{T}\alpha_2 &= \alpha_2, \\ \mathbf{T}\alpha_3 &= -2\alpha_3. \end{aligned}$

(ii)  $B = \begin{pmatrix} 3 & 0 & 0 \\ 0 & 1 & 0 \\ 0 & 0 & -2 \end{pmatrix}.$

(iv)  If $\xi = x\alpha_1 + y\alpha_2 + z\alpha_3$,
then $\mathbf{T}\xi = 3x\alpha_1 + y\alpha_2 - 2z\alpha_3.$

3. (i)  $A = \begin{pmatrix} -1 & -1 \\ -1 & 1 \end{pmatrix}.$

(iii)  $X = \begin{pmatrix} -1 \\ 1 \end{pmatrix}, \quad Y = \begin{pmatrix} -1 \\ 2 \end{pmatrix}.$

(ii)  $C = \begin{pmatrix} -2 & -1 \\ 2 & 2 \end{pmatrix}.$

(iv)  $AX = \begin{pmatrix} 0 \\ 2 \end{pmatrix}, \quad CY = \begin{pmatrix} 0 \\ 2 \end{pmatrix}.$

4. (i)  $B = \begin{pmatrix} 1 & -1 \\ 2 & 1 \end{pmatrix}.$

(iii)  $E = \dfrac{1}{3}\begin{pmatrix} 1 & 1 \\ -2 & 1 \end{pmatrix}.$

(ii)  $D = \begin{pmatrix} 1 & -1 \\ 2 & 1 \end{pmatrix}.$

(iv)  $BE = I = EB.$

(v)  $\begin{aligned} S(4\alpha_1 - 3\alpha_2) &= 7\alpha_1 + 5\alpha_2 \\ &= 2\epsilon_1 + 19\epsilon_2. \end{aligned}$

5. (i)  $F = \begin{pmatrix} 0 & -2 \\ -3 & -1 \end{pmatrix} = BA$     (iv)  $ST\xi = -2y\epsilon_1 - (3x + y)\epsilon_2$ and $TS\xi = -3x\epsilon_1 + (x + 2y)\epsilon_2$.

(iii)  $AB = \begin{pmatrix} -3 & 0 \\ 1 & 2 \end{pmatrix}.$

6. If **T** has rank $r$, the matrix has 1 in each of the first $r$ positions on the diagonal; all other entries are 0.

7. The matrix has 1 in each position immediately below the main diagonal; all other entries are 0.

**EXERCISES 3.5** *pages 145–147*

1. $AB = \begin{pmatrix} 2 & -4 & 3 \\ 2 & 9 & -1 \end{pmatrix}$,   $AC = \begin{pmatrix} -3 & 1 \\ 7 & 3 \end{pmatrix}$,

$B^2 = \begin{pmatrix} 3 & 10 & 14 \\ 2 & -7 & 2 \\ 3 & 3 & -5 \end{pmatrix}$,   $BC = \begin{pmatrix} 3 & 8 \\ -6 & 0 \\ 0 & 1 \end{pmatrix}$,   $CA = \begin{pmatrix} 0 & 4 & 6 \\ -1 & 0 & 1 \\ 3 & 2 & 0 \end{pmatrix}.$

2. The $(i, j)$ entry of $I_m A$ is $\sum \delta_{ik} a_{kj} = a_{ij}$.

The $(i, j)$ entry of $A I_n$ is $\sum a_{ik} \delta_{kj} = a_{ij}$.

3. $A$ is a reflection across the $x$ axis, $B$ is a reflection across $y = x$, and $C$ is a projection vertically onto the line $y = 2x$.

$AB = \begin{pmatrix} 0 & 1 \\ -1 & 0 \end{pmatrix}$, a clockwise rotation of 90°.

$BA = \begin{pmatrix} 0 & -1 \\ 1 & 0 \end{pmatrix}$, a counterclockwise rotation of 90°.

$A^2 = I = B^2$, the identity mapping.

$C^2 = C$, a projection vertically onto $y = 2x$.

4. (ii)  $A = \begin{pmatrix} 1 & 1 & 0 \\ -2 & 1 & 1 \\ 0 & 1 & -1 \end{pmatrix}.$

(iii)  $B = \frac{1}{4}\begin{pmatrix} 2 & -1 & -1 \\ 2 & 1 & 1 \\ 2 & 1 & -3 \end{pmatrix}.$

(iv)  $AB = I = BA.$

5. $\mathbf{T} \rightarrow A = \begin{pmatrix} 0 & 1 & 0 \\ 1 & 0 & 0 \\ 0 & 0 & 1 \end{pmatrix}$, $\mathbf{S} \rightarrow B = \begin{pmatrix} 0 & 0 & 0 \\ 0 & 1 & 0 \\ 0 & 0 & 1 \end{pmatrix}.$

6. Let $a \in F$ correspond to the one-by-one matrix $(a)$.

7. Let $a + ib \in C$ correspond to $\begin{pmatrix} a & b \\ -b & a \end{pmatrix}$.

8. Verify that $N^4 = Z$, but $N^3 \neq Z$.

9. Verify that $A^2 = A$.

11. (ii) $\mathcal{N}(\mathbf{T}_2) = \mathcal{P}_1$; $\mathcal{R}(\mathbf{T}_2) = \{p \in \mathcal{P}_n | p(0) = 0 = p'(0)\}$.

$n(\mathbf{T}_2) = 2$; $r(\mathbf{T}_2) = n - 1$.

(iii) $A_1$ has $0, 1, \ldots, n$ successively down the diagonal; 0 elsewhere. $A_2$ has $(k-1)(k-2)$ in diagonal positions, $k = 1, \ldots, n+1$; 0 elsewhere.

(iv) $\mathbf{T}_1{}^2(p(x)) = \mathbf{T}_1(xp'(x)) = x(xp''(x) + p'(x)) = \mathbf{T}_2(p(x)) + \mathbf{T}_1(p(x))$.

(v) $A_1{}^2$ has $0, 1, 4, \ldots, n^2$ along the diagonal; 0 elsewhere. So does $A_1 + A_2$.

**EXERCISES 3.6** *pages 151–152*

1. (i) $U_{ij}A$ has zeros along row $k$ if $k \neq i$. Row $i$ is $(a_{j1} \ldots a_{jn})$.

2. If $A$ commutes with every $n$-by-$n$ matrix, $A$ must commute with *each* $U_{ij}$. But it commutes with a fixed $U_{ij}$ if and only if $a_{ii} = a_{jj}$, $a_{pi} = 0$ whenever $p \neq i$, and $a_{jq} = 0$ whenever $q \neq j$.

3. Let $D = \text{diag}(d_1, \ldots, d_n)$. $DA$ is obtained by multiplying row $i$ of $A$ by $d_i$, $i = 1, \ldots, n$. Obtain $AD$ by multiplying column $j$ of $A$ by $d_j$, $j = 1, \ldots, n$. If $AD = DA$, $d_i a_{ij} = d_j a_{ij}$ whenever $i \neq j$. For this to hold for all $D$, $a_{ij} = 0$.

4. $A = \begin{pmatrix} 0 & 1 & 0 & 0 \\ 0 & 0 & 2 & 0 \\ 0 & 0 & 0 & 3 \\ 0 & 0 & 0 & 0 \end{pmatrix}$. $A$ is nilpotent of index 4.

5. In the product of $k$ strictly lower triangular matrices, nonzero elements can occur on or below the $k$th subdiagonal.

6–8. Use the isomorphism theorem.

9. The given matrix is singular if and only if its column vectors are linearly dependent, which occurs if and only if $x = 3$.

10. Yes. The conditions are $x - 5y + 10 = 0$, $x + y + 4 = 0$, $x^2 + y^2 + 10 = 36$. The only solution is $x = -5$, $y = 1$.

11. The conditions for $\begin{pmatrix} a & b \\ c & d \end{pmatrix}$ to be idempotent are $a^2 + bc = a$, $b(a+d) = b$,

$c(a + d) = c$, $bc + d^2 = d$. In addition to $I$ and $Z$ the solutions are $\begin{pmatrix} a & b \\ \dfrac{a(1-a)}{b} & 1-a \end{pmatrix}$

and $\begin{pmatrix} a & \dfrac{a(1-a)}{c} \\ c & 1-a \end{pmatrix}$ for any $a$ and any nonzero $b$, $c$.

12. $\begin{pmatrix} ab & b^2 \\ -a^2 & -ab \end{pmatrix}$ for any $a$, $b$.

13. $\begin{pmatrix} a & b \\ c & d \end{pmatrix}$ where $ad \neq bc$.

14. The conditions for orthogonality are $a^2 + c^2 = 1$, $b^2 + d^2 = 1$, $ab + cd = 0$. Suitable combinations of these equations lead to $a^2 + b^2 = 1$, so $c = \pm b$. Similarly $d = \pm a$. But if $c = b$, $d = -a$, and if $c = -b$, $d = a$.

Hence we obtain $\begin{pmatrix} a & -c \\ c & a \end{pmatrix}$ and $\begin{pmatrix} a & c \\ c & -a \end{pmatrix}$, where $a^2 + c^2 = 1$.

Let $a = \cos \Psi$ and $c = \sin \Psi$. This yields

$$R(\Psi) = \begin{pmatrix} \cos \Psi & -\sin \Psi \\ \sin \Psi & \cos \Psi \end{pmatrix} \quad \text{and} \quad S(\Psi) = \begin{pmatrix} \cos \Psi & \sin \Psi \\ \sin \Psi & -\cos \Psi \end{pmatrix}.$$

$R(\Psi)$ is recognized as a rotation of the plane counterclockwise through angle $\Psi$. Also $S(\Psi) = TR(\Psi)$ where $T = \begin{pmatrix} 1 & 0 \\ 0 & -1 \end{pmatrix}$. But $T$ is recognized as a reflection in the $x$ axis. Hence any rigid motion of the plane is a rotation, or a rotation followed by a reflection.

# 4

**EXERCISES 4.1** *pages 158–159*

1. (i) $x_1 = -1 - 3x_4 + x_5$,
    $x_2 = -1 - 3x_4$,
    $x_3 = 1 + x_4 - 2x_5$,
    $x_4$ and $x_5$ arbitrary.

   (ii) $x_1 = -1$,
    $x_2 = -2$,
    $x_3 = 4$.

   (iii) No solution exists.

   (iv) $x_1 = -1 + x_4$,
    $x_2 = 2 - x_4$,
    $x_3 = -x_4$,
    $x_4$ arbitrary.

   (v) $x_1 = 10 + 8x_4$,
    $x_2 = -2 - 2x_4$,
    $x_3 = 0$,
    $x_4$ arbitrary.

2. Same as Exercise 1.

3. (i) $\begin{pmatrix} 1 & 0 & 0 & \frac{1}{6} \\ 0 & 1 & 0 & -\frac{3}{2} \\ 0 & 0 & 1 & \frac{4}{3} \end{pmatrix}.$     (iv) $\begin{pmatrix} 1 & 0 & -1 \\ 0 & 1 & 1 \\ 0 & 0 & 0 \end{pmatrix}.$

(ii) $\begin{pmatrix} 1 & 0 & 0 & 0 & -1 \\ 0 & 1 & 0 & 0 & 0 \\ 0 & 0 & 1 & 0 & \frac{2}{7} \\ 0 & 0 & 0 & 1 & \frac{4}{7} \\ 0 & 0 & 0 & 0 & 0 \end{pmatrix}.$     (v) $\begin{pmatrix} 1 & 0 & -1 \\ 0 & 1 & 1 \\ 0 & 0 & 0 \end{pmatrix}.$

(iii) $\begin{pmatrix} 1 & 0 & 1 \\ 0 & 1 & 1 \\ 0 & 0 & 0 \end{pmatrix}.$

**EXERCISES 4.2** *pages 163–164*

1. (i) $r(A) = 3$.
   (ii) $r(A) = 2$.
   (iii) $r(A) = 4$.

2. By definition all rows of $U_{hi}$ and all columns of $U_{jk}$ are zero except row $h$ and column $k$. The inner product of row $h$ and column $k$ is 0 if $i \neq j$ and 1 if $i = j$.

3. Use matrix multiplication to show that $P_{ij}A$ transposes row $i$ and row $j$ of $A$. Use $M_i(c) = I + (c - 1)U_{ii}$ to show that $M_i(c)A$ multiplies row $i$ of $A$ by $c$. Use either of those methods to show that $A_{ij}A$ adds row $i$ of $A$ to row $j$.

4. Adapt the methods suggested for Exercise 3.

5. Apply the results of Exercises 2, 3, or 4.

6. $M_i(c^{-1})A_{ij}M_i(c)$.

7. $P_{ij} = M_i(-1)A_{ij}M_j(-1)A_{ji}M_j(-1)A_{ij}$.

**EXERCISES 4.3** *page 168*

1. (i) $r(A) = 3$;  $A^{-1} = \frac{1}{8}\begin{pmatrix} -2 & 6 & 4 \\ 1 & -3 & 2 \\ 1 & 5 & 2 \end{pmatrix}.$

(ii) $r(B) = 4$;  $B^{-1} = \dfrac{1}{2} \begin{pmatrix} 1 & 1 & 1 & 0 \\ 0 & 1 & 0 & 1 \\ 1 & 1 & -1 & 0 \\ 0 & 1 & 0 & -1 \end{pmatrix}$.

(iii) $r(C) = 2$.

(iv) $r(D) = 3$;  $D^{-1} = \begin{pmatrix} 2 & 2 & 1 \\ -3 & -3 & -1 \\ 1 & 2 & 2 \end{pmatrix}$.

(v) $r(E) = 3$;  $E^{-1} = \dfrac{1}{2} \begin{pmatrix} 4 & 5 & 8 \\ -2 & -2 & -2 \\ 2 & 3 & 4 \end{pmatrix}$.

2. $P_{ij}{}^t = P_{ij}$; $A_{ij}{}^t = A_{ji}$; $(M_i(c))^t = M_i(c)$.

3. To perform an elementary column operation on $A$, we can perform an elementary row operation on $A^t$ and compute the transpose of the result. Thus $(P_{ij}A^t)^t = AP_{ij}{}^t = AP_{ij}$; $(M_i(c)A^t)^t = A(M_i(c))^t = AM_i(c)$; $(A_{ij}A^t)^t = AA_{ij}{}^t = AA_{ji}$.

4. $I = AA^{-1} = (AA^{-1})^t = (A^{-1})^t A^t$, so $(A^t)^{-1} = (A^{-1})^t$.

**EXERCISES 4.4** *pages 172–173*

1. Reduce each to reduced echelon form. Since these two forms are not identical, $A$ and $B$ are not equivalent.

2. (i) $A$ and $B$ have the same rank and therefore are equivalent.

   (ii) $A, D$, and $E$ are equivalent.

   (iii) The last three are equivalent.

3. (i) $A^{-1} = \dfrac{1}{8} \begin{pmatrix} 2 & 6 & 4 \\ 1 & -3 & 2 \\ 1 & 5 & 2 \end{pmatrix}$.       (iii) $E^{-1} = \dfrac{1}{2} \begin{pmatrix} 4 & 5 & 8 \\ -2 & -2 & -2 \\ 2 & 3 & 4 \end{pmatrix}$.

   (ii) $D^{-1} = \begin{pmatrix} 2 & 2 & 1 \\ -3 & -3 & -1 \\ 1 & 2 & 2 \end{pmatrix}$.

5. From Theorem 4.12 there is one equivalence class for each possible rank. For $m$-by-$n$ matrices, the number of classes is $1 + \min(m, n)$.

6. If $A$ and $B$ are equivalent, so are $A^t$ and $B^t$. But $A^2$ and $B^2$ are not necessarily equivalent, nor are $AB$ and $BA$.

7. Show that $(PAP^t)^t = PAP^t$ if $A$ is symmetric.

8. (i) For each type of elementary matrix $EA$ performs an elementary operation on the rows of $A$, and $AE^t$ performs the same elementary operation on the corresponding columns of $A$.

(ii) If $a_{ii} \neq 0$, $P_{i1}AP_{i1}$ has $a_{ii}$ in the (1, 1) position. If all $a_{jj}=0$ but $a_{ij} \neq 0$, then $A_{ij}AA_{ji}$ has $2a_{ij}$ in the $(j, j)$ position, so $B = P_{j1}(A_{ij}AA_{ji})P_{j1}$ has $2a_{ij}$ in the (1, 1) position.

(iii) From (ii), $b_{11} \neq 0$. Then $M_1(|b_{11}|^{-1/2})BM_1(|b_{11}|^{-1/2})$ has 1 in the (1, 1) position if $b_{11} > 0$, and $-1$ if $b_{11} < 0$. Further row operations and corresponding column operations will produce 0 in the other positions in row 1 and column 1.

(iv) The process can be repeated on the matrix of rows 2 to $n$ and columns 2 to $n$. Hence $QAQ^t$ has $1, -1$, or 0 in each diagonal position and zeros elsewhere. By further permutations of the diagonal elements, the diagonal elements can be arranged so that each diagonal 1 precedes each diagonal $-1$, and so that the latter precede each diagonal 0. Hence a nonsingular $P$ exists so that $PAP^t$ has the block diagonal form

$$\begin{pmatrix} I_p & & \\ & -I_{r-p} & \\ & & Z \end{pmatrix}, \qquad \text{where } r = r(A).$$

## EXERCISES 4.5 *page 176*

| 1. | $m$ | $n$ | $r(A)$ | $r(A|Y)$ |
|---|---|---|---|---|
| (i) | 4 | 5 | 3 | 3, |
| (ii) | 5 | 3 | 3 | 3, |
| (iii) | 3 | 3 | 2 | 3, |
| (iv) | 4 | 4 | 3 | 3, |
| (v) | 3 | 4 | 3 | 3. |

2. (ii) $r(\mathbf{f}) = 1$, $n(\mathbf{f}) = n - 1$.

(iii) The equation determines the null space of a linear functional, a subspace of $R_n$ of dimension $n - 1$. This is a line in $R_2$, a plane in $R_3$, and so on.

3. Reduce the system to echelon form. If $r(A) = r(A|Y) = n$, a unique solution exists. If $r(A|Y) > n$, the system is inconsistent. Geometrically the range space of $\mathbf{T}$, corresponding to $A$, is an $n$-dimensional subspace $\mathscr{R}(\mathbf{T})$ of $R_m$. $\mathbf{T}$ is one-to-one onto $\mathscr{R}(\mathbf{T})$, so a solution exists and is unique if and only if $Y \in \mathscr{R}(\mathbf{T})$.

4. $r(A) = m$.

5. Geometrically, each equation determines a translated plane in $R_3$. The intersection of two such planes is void if $r(A) < r(A|Y)$. If the planes intersect, their intersection is a line when $r(A) = 2 = r(A|Y)$, or a plane when $r(A) = 1 = r(A|Y)$. If a third equation is added, the same geometric description of solutions is valid, and a single point is a solution if $r(A) = 3 = r(A|Y)$.

6. As in Exercise 5, if $m < n$ then $r(A) < n$. If $r(A) < r(A|Y)$ there are no solutions. If one solution exists, then $r(A) = r(A|Y)$, and the solutions form a translated subspace of dimension $n - r(A)$.

**EXERCISES 4.6** *pages 181–182*

1. Let $\alpha = (a, c)$ and $\beta = (b, d)$. The area $A$ of the parallelogram is

$$\|\alpha\| \, \|\beta\| \, |\sin \Psi|,$$

so

$$A^2 = \|\alpha\|^2\|\beta\|^2 - \|\alpha\|^2\|\beta\|^2 \cos^2\Psi$$
$$= (\alpha \cdot \alpha)(\beta \cdot \beta) - (\alpha \cdot \beta)^2.$$

(Incidentally, observe the relation of this equation to the Schwarz inequality.) Computation now yields $A^2 = (ad - bc)^2$.

2. $\det A = (x - 1)(x - 2)(x - 3)$.

3. Expand the determinant and identify the resulting equation. In $E_3$ an analogous four-by-four determinant yields an equation of the plane through three given points.

4. Any elementary column operation multiplies the determinant by a non-zero scalar. But if the columns are linearly dependent, a suitable sequence of elementary column operations produces a matrix with a zero column.

5. Each interchange of two columns changes the sign of a determinant. Consider matrices $B$ and $C$ in which the columns of $B$ are a permutation of the columns of $C$. If a sequence of $k_1$ interchanges of pairs of columns converts $B$ to $C$,

$$\det C = (-1)^{k_1}\det B.$$

If another sequence of $k_2$ interchanges of pairs of columns also converts $B$ to $C$,

$$\det C = (-1)^{k_2}\det B.$$

Hence $k_1 - k_2$ must be even.

**EXERCISES 4.7** *pages 187–188*

1. Apply Theorem 4.17 and note that $a_{11}a_{22} \ldots a_{nn}$ is the only nonzero term.

2. Use the calculations performed in the proof of Theorem 4.17.

3. If $A$ is nonsingular, $AA^{-1} = I$, so $(\det A)(\det A^{-1}) = 1$. If $A$ is singular, a suitable sequence of elementary column operations on $A$ produces a matrix $C$ having a zero column.

4. $\det M_i(c) = c$,   $\det A_{ij} = 1$,   $\det P_{ij} = -1$.

5. If $A$ is singular, so is $A^t$, and $\det A = 0 = \det A^t$. If $A$ is nonsingular, $A$ is the product of elementary matrices, and $A^t$ is the product of the transposes of those elementary matrices, in reversed order.

6. $\sum a_{1j}\text{cof } a_{2j} = (1)(-1)(4-x) + (-1)(1)(x-4) + (2)(-1)(0) = 0.$

$\sum a_{2j}\text{cof } a_{2j} = (-2)(-1)(4-x) + (3)(1)(x-4) + (1)(-1)(0) = x-4 = \det A.$

7. If $k = j$ the Laplace expansion (4.3) yields $\sum a_{ij}\text{cof } a_{ij} = \det A$. If $k \neq j$, $\sum a_{ij}\text{cof } a_{ik}$ is the Laplace expansion of $\det B$, where the columns of $B$ coincide with those of $A$ except that column $k$ of $B$ coincides with column $j$ of $A$. Since $B$ has two identical columns, $\det B = 0$. The second statement of Theorem 4.22 is the transposed form of the first, and $\det A = \det A^t$.

8. $-306$.

9. In the expression for $\det A$ as the sum of signed products of entries of $A$, one from each row and each column, the nonzero products must be of the form

$$(-1)^q b_{p(1),1} \cdots b_{p(m),m} d_{p(m+1),m+1} \cdots d_{p(n),n},$$

where $p$ permutes $\{1, \ldots, m\}$ and $\{m+1, \ldots, n\}$ separately. In the product $(\det B)(\det D)$ the corresponding term is

$$[(-1)^r b_{p(1),1} \cdots b_{p(m),m}][(-1)^s d_{p(m+1),m+1} \cdots d_{p(n),n}].$$

Since $p$ permutes separately the two sets of indices, $(-1)^r(-1)^s = (-1)^q$. Thus the individual terms of the expansion of $\det A$ coincide with the individual terms of the product of the expansions of $\det B$ and $\det D$.

**EXERCISES 4.8** *pages 191–192*

1. $x = 2, y = \frac{3}{2}, z = -\frac{1}{2}.$

2. $x = 2, y = -2, z = -1.$

3. (i) Factor $c$ from each column of $\det (cA)$.
   (ii) $\det A = \det (A^t) = \det (-A) = (-1)^n \det A = -\det A.$

4. Use Theorem 4.22 to show that $A$ com $A = (\det A)I$, whether $A$ is singular or nonsingular. Then calculate the determinant of each side, and argue carefully the singular and nonsingular cases.

5. For general $n$, det $V_n$ is a polynomial $V(x_1, \ldots, x_n)$ of degree $n-1$ in $x_1, \ldots, x_n$ and has the value 0 if $x_i = x_j$ for $i \neq j$. Hence

$$\det V_n = k_n \, \Pi_{1 \leq i < j \leq n}(x_j - x_i), \qquad \text{where } k_n \text{ is a constant.}$$

The coefficient of $x_n{}^{n-1}$ in this expression is $k_n \, \Pi_{1 \leq i < j < n}(x_j - x_i)$. But from the Laplace expansion of det $V_n$, this coefficient is det $V_{n-1}$. Hence

$$k_n \, \Pi_{1 \leq i < j < n}(x_j - x_i) = \det V_{n-1} = k_{n-1} \, \Pi_{1 \leq i < j < n}(x_j - x_i).$$

Hence $k_n = k_{n-1}$. But from (i), $k_2 = 1$, so $k_n = 1$ for all $n$.

6. det $A^t = v_n(1, 2, \ldots, n) \neq 0$.

7. (i)   $W(x) = (b - a)e^{(a+b)x}$; linearly independent.

  (ii)  $W(x) = -b$; linearly independent.

  (iii) $W(x) = 0$; linearly dependent.

  (iv)  $W(x) = 0$; linearly dependent.

8. $W(x) = 2x^3 \begin{pmatrix} 1 & \operatorname{sgn} x \\ 1 & \operatorname{sgn} x \end{pmatrix} = 0$. But if $ax^2 + bx|x|$ is identically zero on $[-1, 1]$, then $a = b = 0$, so the functions are linearly independent.

9. The Jacobian of the transformation $x = g(y)$ is $J = \det (g'(y))$. The two formulas are consistent because if $g$ is increasing, $|J| = g'(y)$; if $g$ is decreasing, $|J| = -g'(y)$, but another change of sign is provided by reversing the limits, since in this case $c > d$ if $a = g(c) < g(d) = b$.

10. Show that $|J| = r$.

11. $\displaystyle \iint_X \sqrt{x^2 + y^2} \, dxdy = 4 \iint_U (u^2 + v^2)^2 dudv.$

# 5

**EXERCISES 5.1** *pages 197-199*

3. For example, $I$ and $2I$.

4. $C = -\dfrac{1}{3}\begin{pmatrix} 6 & 5 & 5 \\ 0 & 4 & 1 \end{pmatrix}$, $P = \begin{pmatrix} 1 & 1 \\ -2 & 1 \end{pmatrix}$, $Q = \begin{pmatrix} 1 & 0 & 1 \\ 1 & 1 & 0 \\ 1 & 1 & 1 \end{pmatrix}$.

5. $B = \begin{pmatrix} 2 & 0 \\ 0 & 0 \end{pmatrix}$, $P = \begin{pmatrix} 1 & 1 \\ 1 & -1 \end{pmatrix}$.

6. (ii)  $C = \begin{pmatrix} 2 & 0 & 0 \\ 1 & 2 & 0 \\ 0 & 0 & 0 \end{pmatrix}$.   (iii)  $Q = \begin{pmatrix} 1 & 1 & 1 \\ 1 & 0 & -1 \\ 0 & 1 & 1 \end{pmatrix}$.

7. Geometrically, an idempotent matrix of rank $k$ represents a projection **T** onto a subspace $\mathscr{R}(\mathbf{T})$ of dimension $k$. Relative to a basis as suggested by Theorem 3.12, **T** is represented by $\begin{pmatrix} I_k & Z \\ Z & Z \end{pmatrix}$.

8. Since the rank of $BA$ need not be the rank of $AB$, these matrices are not necessarily similar. But if $A$ is nonsingular, $BA = A^{-1}(AB)A$.

9. In each case the given matrices are similar.

11. $P^{-1} = \dfrac{1}{3}\begin{pmatrix} 1 & 1 & 1 \\ 1 & e^2 & -(1+e^2) \\ 1 & -(1+e^2) & e^2 \end{pmatrix}$.

12. Let $S(\gamma_i) = \alpha_i$. Recalling that $T_1(\alpha_j) = \sum a_{ij}\alpha_i$ and $T_2(\gamma_j) = \sum a_{ij}\gamma_j$, show that $ST_2(\gamma_j) = T_1 S(\gamma_j)$.

**EXERCISES 5.2** *pages 205–207*

1. (i) $\lambda_1 = 2$, $\lambda_2 = -3$; $X_1 = a\begin{pmatrix} 1 \\ 1 \end{pmatrix}$, $X_2 = b\begin{pmatrix} 2 \\ -3 \end{pmatrix}$; let $a = b = 1$.

(ii) $\lambda_1 = -1 = \lambda_2$, $\lambda_3 = 8$; $X_1 = a\begin{pmatrix} 1 \\ 0 \\ -1 \end{pmatrix} + b\begin{pmatrix} 0 \\ 2 \\ 1 \end{pmatrix}$, $X_3 = c\begin{pmatrix} 2 \\ 1 \\ 2 \end{pmatrix}$.

Let $a = 1$, $b = 0$; and then let $a = 0$, $b = 1$; let $c = 1$;

(iii) $\lambda_1 = 1 = \lambda_2 = \lambda_3$, $\lambda_4 = 3$; $X_1 = a\begin{pmatrix} 1 \\ 0 \\ -2 \\ 0 \end{pmatrix}$, $X_4 = b\begin{pmatrix} 1 \\ 0 \\ 0 \\ 0 \end{pmatrix}$.

2. (i) The characteristic polynomial is $-(\lambda + 2)(\lambda - 1)(\lambda - 3)$.

3. The diagonal entries are the characteristic values.

4. If $AX = \lambda X$, then $\lambda^{-1}X = A^{-1}X$.

5. Show that if $AX = \lambda X$, then $A^k X = \lambda^k X$, so that for any polynomial $p$, $p(A)X = p(\lambda)X$.

6. (i) $\lambda = 0$ and $\lambda = 1$.

(ii) $\lambda = 0$.

(iii) Any nonzero scalar.

7. Note that $p(0) = \lambda_1 \lambda_2 \ldots \lambda_n$ and also $p(0) = \det A$.

8. (i) For the inductive step expand $\det(C - \lambda I)$ by the elements of the first column.

9. Use induction on $k$.

10. Show that $\mathscr{C}_0$ is closed under the vector space operations. Then for each $\xi \in \mathscr{C}_0$, $\mathbf{T}\xi = \lambda_0\xi \in \mathscr{C}_0$, so $\mathscr{C}_0$ is T-invariant.

11. Let $\alpha \in \mathscr{C}_1 \cap (\mathscr{C}_2 + \ldots + \mathscr{C}_k)$. Then $\mathbf{T}\alpha = \lambda_1\alpha = \lambda_2\xi_2 + \ldots + \lambda_k\xi_k$, since $\alpha = \xi_2 + \ldots + \xi_k$ for suitable $\xi_i \in \mathscr{C}_i$. By Exercise 9 $\{\alpha, \xi_2, \ldots, \xi_k\}$ is linearly independent, since the characteristic values are distinct. Thus the equation $(\lambda_2 - \lambda_1)\xi_2 + \ldots + (\lambda_k - \lambda_1)\xi_k = \theta$ implies that $\xi_2 = \xi_3 = \ldots = \xi_k = \theta$. Hence $\alpha = \theta$.

## EXERCISES 5.3 *pages 213–214*

1. In this exercise there are various choices of characteristic vectors and thus various choices of diagonalizing matrix $P$. Solutions different from those given below quite possibly are correct.

(i) $\lambda_1 = 1 = \lambda_2 = \lambda_3$;  $X_1 = a\begin{pmatrix}1\\1\\0\end{pmatrix} + b\begin{pmatrix}-1\\0\\3\end{pmatrix}$;  not diagonable.

(ii) $\lambda_1 = 3 = \lambda_2$;  $\lambda_3 = 12$;  $P = \begin{pmatrix}1 & 1 & -4\\-1 & 0 & -4\\0 & 1 & 1\end{pmatrix}$.

(iii) $\lambda_1 = 1 = \lambda_2$;  $\lambda_3 = 2$;  $X_1 = a\begin{pmatrix}5\\2\\-5\end{pmatrix}$, $X_3 = b\begin{pmatrix}1\\1\\-2\end{pmatrix}$; not diagonable.

(iv) $\lambda_1 = 1$,  $\lambda_2 = -1$,  $\lambda_3 = -2$;  $P = \begin{pmatrix}1 & 0 & 0\\1 & 2 & 1\\0 & 1 & 1\end{pmatrix}$.

(v) $\lambda_1 = 1 = \lambda_2$;  $\lambda_3 = -2$;  $P = \begin{pmatrix}1 & 0 & 1\\0 & 0 & -1\\0 & 1 & 2\end{pmatrix}$.

(vi) $\lambda_1 = 1 = \lambda_2$;  $\lambda_3 = 5$;  $P = \begin{pmatrix}-2 & -1 & 1\\1 & 0 & 1\\0 & 1 & 1\end{pmatrix}$.

2. (i) Diagonable, since the characteristic values are distinct.

(ii) Diagonable, since there are three linearly independent characteristic vectors.

(iii) Not diagonable, since there are not four linearly independent characteristic vectors.

3. (i) Diagonable; the double root has two associated characteristic vectors that are linearly independent.

(ii) Not diagonable, since the triple root determines a one-dimensional characteristic subspace.

(iii) Not diagonable; the double root determines only a one-dimensional characteristic subspace.

(iv) Diagonable; the characteristic values are distinct.

4. $A$ is diagonable, with $n$ as a simple characteristic value and 0 as an $(n-1)$-fold characteristic value with $n-1$ linearly independent characteristic vectors. Hence $B$ is the diagonal matrix to which $A$ is similar.

5. (i) $E_i = E_i(E_1 + \ldots + E_n) = E_i^2$.

(ii) $F_i^2 = (PE_iP^{-1})(PE_iP^{-1}) = PE_i^2P^{-1} = F_i$.

$F_iF_j = (PE_iP^{-1})(PE_jP^{-1}) = Z$ if $i \neq j$.

$\sum F_i = \sum PE_iP^{-1} = P\sum E_iP^{-1} = I$.

6. Various correct spectral forms exist.

For
$$A = \begin{pmatrix} 7 & 4 & -4 \\ 4 & 7 & -4 \\ -1 & -1 & 4 \end{pmatrix}, \quad \text{let } P = \begin{pmatrix} 1 & 0 & 4 \\ -1 & 1 & 4 \\ 0 & 1 & -1 \end{pmatrix}.$$

Then $P^{-1}AP = \operatorname{diag}(3, 3, 12)$. Let $F_1 = PE_1P^{-1}$ and $F_2 = PE_2P^{-1}$, where $E_1 = \operatorname{diag}(1, 1, 0)$ and $E_2 = \operatorname{diag}(0, 0, 1)$. Then $A = 3F_1 + 12F_2$.

For
$$A = \begin{pmatrix} 2 & 0 & 0 \\ 0 & 1 & 0 \\ 0 & 2 & 2 \end{pmatrix}, \quad \text{let } P = \begin{pmatrix} 1 & 0 & 0 \\ 0 & 0 & 1 \\ 0 & 1 & -2 \end{pmatrix}.$$

Then $P^{-1}AP = \operatorname{diag}(2, 2, 1)$. Let $F_1 = PE_1P^{-1}$ and $F_2 = PE_2P^{-1}$, where $E_1$ and $E_2$ are as defined above. Then $A = 2F_1 + F_2$.

7. (i) If $n = 1$, $A = Z$ and $A^2 = A$; if $n = 2$, then by the Cayley-Hamilton Theorem, $A^2 + c_1A = Z$;

(ii) for $n > 2$ let $A = \operatorname{diag}(0, 1, 2, \ldots, n-1)$; then $A^2$ is not proportional to $A$.

8. (i) $A = \begin{pmatrix} a & b \\ -\dfrac{1+a^2}{b} & -a \end{pmatrix}$.

(ii) If $A$ is three-by-three, its characteristic polynomial is a real cubic, which has a real root. Hence $A^2$ has a nonnegative real characteristic value. But the characteristic values of $-I$ are all $-1$, so $A^2 \neq I$.

9. The constant term of the minimum polynomial of $A$ has each characteristic value of $A$ as a factor.

10. (i)  $A$ is not diagonable, so $m(x) = (x - 1)^2(x + 1)$;

   (ii)  $A$ is diagonable, so $m(x) = (x - 3)(x - 12)$,

   (iii)  $A$ is diagonable, so $m(x) = (x - 1)(x - 5)$,

   (iv)  $A$ is not diagonable; $m(x) = (x - 1)^3$.

**EXERCISES 5.4** *pages 220–221*

1.  $q(\xi + \eta) + q(\xi - \eta) = (X + Y)^t A(X + Y) + (X - Y)^t A(X - Y)$
$$= 2X^t AX + 2Y^t AY = 2q(\xi) + 2q(\eta).$$

2. (i)  $A = \begin{pmatrix} 1 & 0 & -1 \\ 0 & 2 & 2 \\ -1 & 2 & 6 \end{pmatrix}$; rank $= 3$, signature $= 3$.

   (ii)  $A = \begin{pmatrix} 0 & 8 & 0 \\ 8 & 0 & 0 \\ 0 & 0 & -1 \end{pmatrix}$; rank $= 3$, signature $= -1$.

   (iii)  $A = \begin{pmatrix} 3 & 2 & 4 \\ 2 & 0 & 2 \\ 4 & 2 & 3 \end{pmatrix}$; rank $= 3$, signature $= -1$.

3.  Let $K$ be skew; since $X^t KX$ is a scalar $c$, $c = (X^t KX)^t = -c$.

4.  If $A^t = \pm A$, $(A^2)^t = (A^t)^2 = A^2$.

7.  $A = \frac{1}{2}(A + A^t) + \frac{1}{2}(A - A^t)$. Calculate each.

8. (i)  $X^t AX = 2x_1^2 + 4x_1 x_2 - 4x_1 x_3 + 2x_2 x_3 + x_3^2$.

   (ii)  $X^t SX = X^t AX$, $X^t KX = 0$.

   (iii)  See Exercise 3.

10.  A quadratic form $q$ on $E_n$ can be represented by a matrix of the block diagonal form

$$F = \begin{pmatrix} I_p & & \\ & -I_{r-p} & \\ & & Z \end{pmatrix}$$

As in the proof of Theorem 5.16, $q$ is positive definite if and only if $p = n = r = s$. But $q$ is also represented by $A$ if and only if $A = Q^t FQ$ for some nonsingular $Q$. $A$ and $F$ have the same rank and signature.

11. (i) $r(A) = 2,\ s(A) = 0,$  (iv) $r(D) = 3,\ s(D) = -1,$

(ii) $r(B) = 2,\ s(B) = 2,$  (v) $r(E) = 2,\ s(E) = 0,$

(iii) $r(C) = 3,\ s(C) = 1,$  (vi) $r(F) = 2,\ s(F) = 0.$

**EXERCISES 5.5** *pages 225–227*

2. (i) Write $ax_1{}^2 + 2bx_1x_2 + cx_2{}^2 = a\left(x_1 + \dfrac{b}{a}x_2\right)^2 + \left(\dfrac{ac - b^2}{a}\right)x_2{}^2.$

(ii) For $a > 0$ the conic is an ellipse when $b^2 - ac < 0$, a parabola when $b^2 - ac = 0$, and a hyperbola when $b^2 - ac > 0$.

3. Rank   Signature   Quadric surface

| | | |
|---|---|---|
| 3 | 3 | ellipsoid, |
| 3 | 1 | hyperboloid – one sheet, |
| 3 | -1 | hyperboloid – two sheets, |
| 2 | 2 | elliptical cylinder, |
| 2 | 0 | hyperbolic cylinder, |
| 1 | 1 | two intersecting planes. |

4. Near $(a, b)$ the value of the quadratic form in brackets is the dominating term in determining the values of $f$. This form must be positive definite if $f(a, b)$ is a relative minimum, and it must be negative definite if $f(a, b)$ is a relative maximum.

5. (i) $1 = \det (PP^{-1}) = \det (PP^t) = (\det P)^2.$

(ii) Let $PX = \lambda X$. Then $\|X\| = \|PX\| = \|\lambda X\| = |\lambda|\ \|X\|$, where $\|X\| \neq 0$.

6. Observe that $(I - K)$ commutes with $(I + K)^{-1}$ since it commutes with $(I + K)$. Let $P = (I - K)(I + K)^{-1}$ and verify that $P^t = P^{-1}$.

# Index

Adjoint of a matrix, 187
Albert, A. A., 52
Alternating function, 180
Angle between vectors, 7–8, 34, 109–110
Area of a parallelogram, 8, 177
Associativity, 10, 63, 120, 122, 144

Basis, 28, 97; for an idempotent transformation, 132, 140; for a nilpotent transformation, 140; normal orthogonal, 36–37, 111–112, 221–225; ordered, 25, 100; orthogonal, 36, 111–112, 221–225; of range space, 124; standard, 29
Bell, E. T., 52
Bessel's inequality, 113
Bilinearity, 35, 50, 63, 120, 122, 144

Canonical form, 170; for congruence, 219; for equivalence, 171; for similarity, 197, 212
Cartesian product, 13
Cartesian equations, 42, 44, 46–48
Cayley-Hamilton Theorem, 211–212, 214
Characteristic: equation 201; polynomial, 201, 205–207, 211, 222; subspace, 204–207; value, 69, 79, 200–201, 205–207, 211–212, 222; vector, 69, 79, 200–201, 205–207, 211, 222–223

Cofactor, 185–187
Column rank, 162
Column vector, 136
Comatrix, 187, 191
Commutativity, 10, 63, 151
Companion matrix, 206
Complement, orthogonal, 113
Complex numbers, 84, 98, 103, 146, 201, 226
Congruence of matrices, 216–225
Conjugate of a complex number, 226
Conjugate-bilinearity, 106
Conjugate-symmetry, 105
Consistency condition, 175
Conte, S. D., 26
Convex linear combination, 43, 77
Convex set, 43, 48
Coordinate system. *See* Basis
Cramer's rule, 189
Cross product, 49–53
Curtis, C. W., 52

Derivative, partial, 47, 226
Derivative operator, 120, 124
Determinants, 50–51, 176–192
Diagonal entries of a matrix, 64, 148
Diagonability, 207–213, 221–225